Management of nitrogen and water in potato production

Management
of
nitrogen and water
in
potato production

Editors: A.J. Haverkort & D.K.L. MacKerron

Wageningen Academic
P u b l i s h e r s

Subject headings:
Potato
Nitrogen and water
Decision support

The publisher is not responsible for possible damages, which could be a result of content derived from this publication.

This book was written and published as one of the activities of a Concerted Action funded by the European Commission, Contract No. CT96 - 1560, -Efficiency in Use of Resources: Optimization in Potato Production (EUROPP).

Layout and figures
Janneke Abbas

ISBN-10: 90-74134-77-7
ISBN-13: 978-90-74134-77-4

The editors and authors gratefully acknowledge the support of the Commission. They also wish to make clear that the contents of this book, the presentation of facts, the deductions, and the opinions that it contains, are entirely their responsibility and in no way represents the view of the Commission or its services.

First published, 2000
Reprint, 2006

Wageningen Academic Publishers
The-Netherlands, 2006

Preface

Management of Nitrogen and Water in Potato Production

The supply of water and nitrogen are key problems in the production of the potato crop. There is a balance between their provision and the returns from them that can strongly influence the profitability of the crop as both the level of yield and the quality of the tubers are critically dependent on the correct supply of both water and nitrogen. Over-supply and deficiency both lead to problems in the crop. Additionally, over-supply leads to environmental problems. In the case of water, it means that a limited resource is used inefficiently and it may lead to leaching. Where too much nitrogen is given, the excess will be leached in to the ground water. It is because of those possible adverse effects that the supply of nitrogen and the extraction of water can each be the subject of regulation by government authorities. This book was produced in the course of an EU-funded Concerted Action into the efficient use of nitrogen and water in the potato crop, called EUROPP – Efficiency in the Use of Resources: Optimisation in Potato Production.

This manual explains the roles of water and nitrogen in the growth and development of the potato crop in Chapter 1 and, in the following chapters, it presents an assessment of the procedures available to achieve the correct balance between input and production. Chapter 2 considers trends in plant nitrogen status (total N and nitrate), how that status can be determined by invasive or non-invasive methods, how it is best sampled, and what is the role of such measurements in supporting decisions on application of nitrogen. Similarly, Chapter 3 addresses questions of the nitrogen status of the soil, considering both inorganic and organic sources of nitrogen, shows the importance of adequate sampling. It explains the importance of nitrification and denitrification and how losses occur. Chapter 4 explains the roles of water in plants and soil, how water status can be measured in each medium, and how those measurements can be used in practice. Chapter 5 reviews the existing decision support systems for the supply of nitrogen and water, which sources of nitrogen may be used (organic manures as well as bagged fertiliser), what irrigation practices are suitable, and which decision support systems are available for water and nitrogen. It also considers the important differences between irrigated and rain-fed potato production. Chapter 6 explores possible future developments: What is the role of crop growth modelling and computer simulation? What are the risks and lessons from trends in current practices? What do we know about the consequences of 'organic' farming? What are the opportunities from precision agriculture? And what do we still need to learn? The book concludes with perspectives on recent developments including those emanating from the concerted action.

The book updates knowledge in a simple, easily understandable way and it is intended for farmers, farmers' consultants, researchers and decision makers at the levels of farming, research, and policy-making.

The editors

Contents

1   Introduction

1.1   The purpose of this Manual of Methods

D.K.L. MacKerron

The supply of water and nitrogen are key problems in the production of the potato crop. Providing these materials to the *crop* costs the farmer money and so there is a balance between the provision of these resources and the returns from them that can strongly influence the profitability of the crop as the level of yield and the quality of the tubers are critically dependent on the correct supply of both water and nitrogen. Over-supply and deficiency both lead to problems in the crop. Additionally, the demands of the crop for those commodities and the provision of them by the farmer can have adverse effects on the environment if they are not managed properly. It is because of those possible adverse effects that the supply of nitrogen and the extraction of water can each be the subject of regulation by government authorities.

Throughout most of Europe, irrigation is necessary for potato crops to perform at or close to the potential levels defined by temperatures and solar radiation and, where irrigation is used, it is important that the farmers should use a recognised method to schedule its application. A number of commercial systems are available for that purpose. The correct level of nitrogen supply is much harder to achieve. There are no readily available systems for recommendations, and the availability of nitrogen is dependent on the water status of the soil. This manual, therefore, sets out briefly to explain the roles of water and nitrogen in the growth and development of the potato crop and then to present, in a simple and uncomplicated way, an assessment of the procedures available to achieve the correct balance between input and production.

The growth and development of the potato crop respond to the supplies of water and nitrogen, on a very wide range of scales, from cell, through whole plant to field-scale, and several of the processes involved interact with each other. For example, the supplies of both water and nitrogen influence early

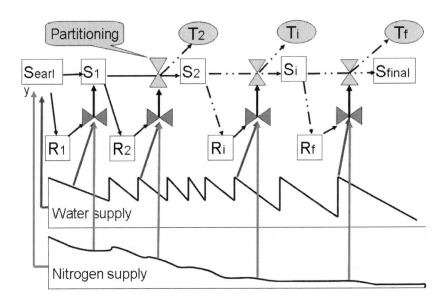

Figure 1   *Schematic indication of interactions between root and shoot growth, the uptake of*
              water and nitrogen, and partitioning to tubers through the growing season. S =
              Shoot, R = Root, T = Tubers; early, 1, 2, i, f = sequence.

shoot growth. All three influence root growth. That, in turn, influences future
supplies of water and nitrogen thereby moderating future shoot growth. Soil
water status influences the availability of mineralized nitrogen, and the nitrogen
status of the whole plant influences crop development and tuber maturity. A
schematic, Figure 1, outlines some of the processes and their interactions
through the growing season. Processes over winter and during the remainder of
the crop rotation have been omitted for simplicity.

The Concerted Action - EUROPP

This book was produced in the course of an EU-funded Concerted Action into
the efficient use of nitrogen and water in the potato crop. The concerted action
was called EUROPP – Efficiency in the Use of Resources: Optimisation in
Potato Production. Why was there such a Concerted Action? There are many
programmes of research on aspects of water supply and nitrogen nutrition for

the potato crop, conducted in almost all of the countries of Europe because these two resources are so very important for the potato crop. The issues are complex, however, and cover a range of scales from plant physiology and the dynamics of soil microbiology to variation in an arable field. EUROPP was planned to provide a link between national programmes, leading to the pooling of data, and harmonising practices. It provided a forum that enabled researchers throughout Europe to collaborate, exchange ideas, and develop common protocols. A principal focus of the Action has been collectively to develop methodologies to be used, as appropriate, by producers, their advisers, and laboratories. This manual is one product of that effort at harmonisation.

There are other, non-agricultural issues to be considered also. Governments, and the population as a whole, are concerned that agricultural practices may adversely affect the wider environment. For example, the rising levels of nitrate in many rivers and in ground waters are blamed on changes in agricultural practices: the ploughing of grassland, injudicious disposal of animal wastes, and simple over-fertilization of some crops. Regulations are now in force in a number of areas to limit the application of nitrogen to land. (For example, the EU Nitrate Directive). Yet high yields and good quality are essential if farmers are to achieve a satisfactory level of profitability and to remain competitive in world markets. There is a clear need for farmers to have a set of methods that will allow them to produce crops at near to potential levels and yet minimize their impact on the environment. Also, the regulatory authorities need to have a recognised set of rational standards by which they can assess farm practices in their areas, to be satisfied that farmers are doing as best they can. This manual is intended to be of some help to regulators as well as to farmers and their advisers.

What is optimization and why do it?

According to the 'Law of Limiting Factors', if a resource such as nitrogen is provided at adequate levels for one set of environmental conditions, it will be either in excess or limiting when those conditions change. Again, the 'Law of Diminishing Returns' states that the additional benefit from higher and higher levels of a resource declines. The corollary of that is that the penalty for being slightly deficient is small at levels of input close to those for maximum production. Now, what has that got to do with water and nitrogen in the potato crop?

Where water and nitrogen are supplied so that the crop will not be deficient in either, then most of the time they will be in excess. The farmer will spend more than he needs to and the excess amount becomes a pollutant. Yet, if insufficient is applied, production is less than potential and income is lost. Between those conditions lies a more-or-less correct application. The eventual requirement of the crop for applied nitrogen cannot be known with certainty as it depends on the growth of the crop and the supply of native nitrogen from the soil. Both are affected by weather. Where the nitrogen supply is optimized, leaf expansion and tuber growth are barely limited and, yet, there is little residual mineral nitrogen in the soil at the end of the growing season. It entails giving enough nitrogen so that the risk of providing an excessive amount balances the risk of the crop being deficient. Where the use of water is optimized in the early season, the expansion of the leaf canopy proceeds at its potential rate but there is no drainage. Later in the season the crop is allowed to use the water stored in the soil reservoir and dry out the soil, but not so much that the crop is stressed and senesces prematurely, or is liable to growth disorders, such as growth cracking if there is rain subsequently.

Although the farmer, generally, supplies water and nitrogen independently, the uptake of water and nitrogen are coupled. The plant takes up nitrogen from the soil solution, in the transpiration stream, and so the issues of water supply and N-supply ought not to be considered separately. Optimizing the supply of both water and nitrogen for production requires that the two resources should be just limiting production, for equal amounts of time.

Here, in this manual, is the opportunity to use the results of recent and current research to match inputs of nitrogen to crop requirements and thereby to maximize efficiency, improve competitiveness of European food production in the world market, reduce demands on resources, and minimize pollution. The potato crop can be grown with optimized inputs of water and nitrogen and there are reasonable standards for farmers to observe and so minimize the effects of their enterprise on the environment. This manual is intended to explain the methods to achieve those things.

1.2   Basic concepts of the management of supply of nitrogen and water in potato production

J. Vos & D.K.L. MacKerron

What is the role of nitrogen and water in potato production? Which plant processes are particularly sensitive to nitrogen and water? What are the natural and man-supplied sources of the growth factors? What are principles of management of nitrogen and water the grower needs to understand and to apply to ensure unrestricted crop growth and minimize environmental pollution?

Introduction

Nitrogen is an essential nutrient in plant growth, as it is an important constituent of the building blocks of almost all plant structures. At sub-cellular level it is essential in amino acids, enzymes, and proteins. Therefore at the macroscopic level it is essential in the production of leaves and tubers. Arable crops including potato get some of their nitrogen supply from 'native' nitrogen from the soil. They may also get some from added manures, if they have been used, and they get the rest from added fertilizer. Unlike the case with most nutrients, an analysis of soil, made some time before planting, is not a sufficient guide to what the soil will supply. This is because nitrogen in the soil is subject to several transformation processes that affect the amount of plant-available nitrogen and, also, the loss of nitrogen from the system. This chapter introduces the dynamics of changing nitrogen supply in the soil as well as the use of nitrogen by the crop plant.

Water, which is essential for plant growth, should be used efficiently. Firstly, because it is a scarce resource and, secondly, because its adequate supply is necessary for the fullest use of the available nutrients in the soil.

The objectives of managing the supply of nitrogen and water to the crop are to ensure production that is close to potential, to optimize the use of all resources and to minimize environmental pollution. This introductory chapter presents the basic concepts and principles of the processes and factors that determine the supply of nitrogen and water to the crop. This chapter provides the common basis for later chapters, each of which deals with particular aspects in detail.

The roles of nitrogen and water in plant functioning

Most plant constituents contain nitrogen, represented by the symbol N. Proteins are particularly rich in N. There are several types of proteins in plant organs including enzymes, which are involved in all biochemical processes of growth and development. In storage organs, such as potato tubers there are storage proteins as well as starch. Together these form the raw material from which the next potato generation is formed and are, also, the food that we eat.

Normally plants get their nitrogen as inorganic nitrogen absorbed by the roots from the soil solution. Two forms of ionic nitrogen can be absorbed, namely ammonium (NH_4^+) and nitrate (NO_3^-). The sum of nitrate and ammonium is called total mineral nitrogen, or just mineral N. Soil tests to assess the amount of nitrogen in the soil that is available to plants need, therefore, to include both ammonium and nitrate although, usually, nitrate dominates over ammonium in soils. Plant leaves, too, have the capacity to absorb nutrients and foliar feeding with nitrogen, often in the form of urea, is based on that capacity.

Depending on the yield, a potato crop can take up in the order of 125 to 250 kg N ha^{-1}. The average nitrogen concentration in dry plant material in young plants can exceed 5%. Thereafter, growth (carbon fixation) occurs faster than the uptake of nitrogen and so the concentration of nitrogen declines, falling to 1.5% or lower at maturity, even in well-fertilized plants. The several tissues of the plant differ in their typical nitrogen concentrations. In green leaves the concentration of nitrogen varies between 7% (ample supply of nitrogen) and 2.5% (deficient in nitrogen). In stems the values range from 6% (ample) to 0.5% (deficient) and in tubers from 3.5% to 0.5% (e.g. Young et al., 1993).

Water has several functions. Principally, all synthetic biological processes occur in solution. Green plant material can be 90% water and potato tubers are 75 to 83% water at maturity.

Although it is important that plants are hydrated, the amount of water in a plant is very small compared to the amount that it loses to the atmosphere in transpiration, the name given to evaporation through leaves (see later). For example, a mature, standing potato crop may contain from 8 – 10 kg of water per square metre, i.e. 8 – 10 litres or the equivalent of 8 – 10 mm of rain water, of which about 75% is in the tubers. Transpiration is unavoidable because the leaves are wet inside, and the air is dry, but it is an acceptable necessity because transport processes in the soil and in the plant depend on transpiration. Water is the solvent carrier for minerals in the soil and for minerals and sugars in the plant. In addition, the expansion of leaves is enabled by water pressure in the cells of the leaves.

Water can evaporate directly from the soil surface. The drier the top soil, and the larger the proportion of the soil that is covered by the crop, the lower will be the rate of direct evaporation and the higher the proportion of evaporation that is transpiration. The total water loss from the soil - plant system is called evapotranspiration, in formula:

Evapotranspiration = evaporation (from soil) + transpiration (from plants).

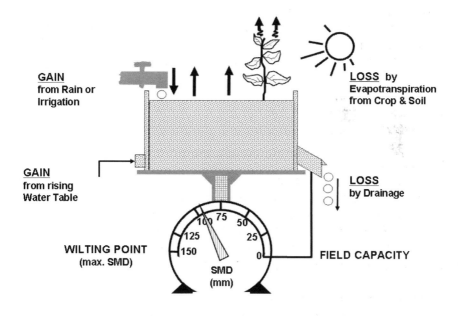

GAIN from Rain or Irrigation

LOSS by Evapotranspiration from Crop & Soil

GAIN from rising Water Table

LOSS by Drainage

WILTING POINT (max. SMD)

100 75 50
125   25
150   0

SMD (mm)

FIELD CAPACITY

Figure 1   A 'water balance'.

Figure 1 illustrates the balance between inputs of water to the soil and losses from it. The soil represents a reservoir of usable water. As the crop draws on that reservoir a soil moisture deficit is developed that is expressed as the amount of rain that will be needed to restore the soil to holding as much water as it can. That state is called 'field capacity'.

Where water is in short supply it is important to minimize water losses by evaporation from the soil, which can be thought of as water wasted. Since evaporation requires energy, transpiration keeps the plant cool and that can be an important function of transpiration.

The rate of evapotranspiration depends on the intensity of solar radiation and on the relative humidity of the air. The more energy is received from the sun and the drier the air, the faster the rate of evapotranspiration will be. Depending on weather conditions (cloudiness, location in Europe, and date) the daily rate of evapotranspiration varies between 1 and 8 mm. The total amount of water that is needed to grow a crop well depends on the weather and on the length of the growing season (and so maturity class of the cultivar), but can be in the order of 300 to 700 mm. The crop can consume in the order of 50 - 60 litres of water per kilogramme of fresh tubers.

Plant processes affected by nitrogen

There are differences in growth and development of the crop where the supply of nitrogen and water are ample, adequate, or low and both the yield and quality of the crop are reduced where there is an under- or over-supply of either. Where the fertiliser applied at planting is adequate for the whole season or not quite adequate, the initial development of the crops will be similar. The differences develop as available nitrogen is used up and the rate of supply to the crop declines from the ideal. Supplementary application of nitrogen can maintain the rate of supply to the crop. The objective of the grower, clearly, is to supply both water and nitrogen at near to the optimal rates.

Potato plants grown on a wide range of nitrogen applications show surprisingly few differing plant attributes. Those few attributes or processes that are very sensitive to nitrogen supply include the degree of branching and leaf size. Greater supply of nitrogen causes an increased level of branching, particularly giving apical branches at the top of the plant resulting in a larger

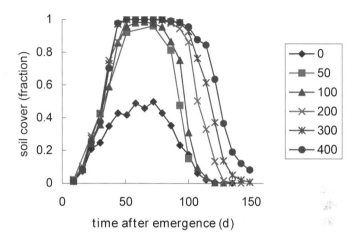

Figure 2   The change in fractional soil cover with time for potatoes receiving 0, 50, 100, 200, 300 or 400 kg N ha⁻¹.

number of leaves per plant. Leaf colour is another property that is sensitive to nitrogen supply: nitrogen-deficient plants have a lighter green colour than plants with abundant N supply. Unfortunately, the corollary does not hold. Leaves that are a dark green may indicate plants that are short of water – and so have plenty of nitrogen for their present state.

Through its effects on branching and sustained leaf production, nitrogen supply affects the potential duration of the growing period and particularly the duration of the period of full soil cover when the daily rate of production is maximal (Fig. 2). The fraction of soil that is covered by leaves indicates the proportion of the solar radiation that is being absorbed by the crop.

The faster full soil cover is reached and the longer that cover lasts, the more radiation is intercepted by the crop and the higher the yield. So, nitrogen supply primarily affects the total amount of solar radiation that is intercepted during the growing season. It has little effect on the amount of plant material produced per unit of absorbed solar radiation (Millard and Marshall, 1986).

Plants grown under N limitation have a lower nitrogen concentration in the dry matter that is already noticeable in the early stages of a crop's life. The decline in concentration of nitrogen in the plant dry matter with growth stage and time has already been described. That change reflects the increasing proportions of structural and storage material in the plant, which have lower concentrations of nitrogen, and a decreasing proportion of metabolic material. The nitrogen concentration expressed in terms of the fresh weight or, better, the water content is less variable through the season and can provide a conservative estimate of the adequacy of nitrogen nutrition (MacKerron et al., 1994).

The plant regulates the rate of nitrogen uptake from the soil. Potato plants that are supplied with more nitrogen than is enough for their immediate needs will take up nitrate at luxury levels (Millard et al., 1986) without metabolising it, but will store the nitrate in the vacuoles until the supply is less than the requirement. N-deficient plants show little or no accumulation of nitrate. Hence, the nitrate concentration is assumed to mirror the nitrogen status of the crop and nitrate tests have been developed based on that assumption.

There can be no single 'correct' value for the level of fertilizer nitrogen required by the potato crop. If the 'optimum' were to be defined simply as the level that gave the highest yield, that level would change progressively through the growing season (Fig. 3) and would depend on the intended date of harvest. In other words, the 'optimum' for a crop to be harvested in early August would be different from that for one to be harvested in mid-September.

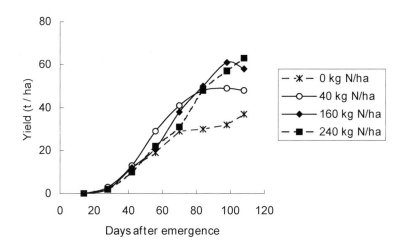

Figure 3   *Development of yields in potatoes grown with differing supplies of nitrogen. (See*
            text for further explanation. Data: MacKerron unpublished).

In Figure 3, an application of only 40 kg N /ha was sufficient to allow fast canopy closure and so there was no penalty in yield until later in the season, when the canopy began to die early (Fig. 2). Plants with moderate rates of supply, sufficient to allow fast canopy closure, had the higher yields until later in the season. The yield of plants given the highest rate of supply (240 kg N / ha) did not catch up with the yield from those given a 'standard' rate of supply until the end of the growing season.

Optima defined in this way will differ from year to year and place to place with soil and weather. That definition is also inadequate because of the diminishing returns from higher inputs. Figure 4 shows how, in a high-yielding, late-cropping situation, tuber yield can appear to increase at ever-higher rates of application. However, the smaller increases were achieved at lower dry matter concentrations and the residual nitrogen increased directly with application rate. The uptake of N seems to increase linearly with increase in application rate (Fig. 4), but careful examination of the response also shows diminishing efficiency of uptake with increase in application rate. It can be inferred from Figure 4 that the concentration of nitrogen in the crop increases steadily with increase in N application rate (cf. MacKerron et al., 1993; Vos, 1997). Potential losses of nitrogen increase with higher application rates too. This is shown by the curve for 'N excess' in Figure 4. N excess being defined by:

$$N_e = N_a - N_u + N_0$$

Where N_e = excess nitrogen, N_a = nitrogen application, N_u = nitrogen uptake by the fertilized crop, and N_0 = nitrogen uptake by a zero N control

Excess nitrogen is the difference between the application rate and the resulting uptake of N, corrected for the uptake that results from sources other than fertilizer application, which is assumed to be equal to the uptake that is achieved by an unfertilized control plot. Part of the N excess will be lost to the environment. Some of the nitrogen that remains in the leaves and stems at harvest may also be lost to the environment. Evidently, emission losses potentially increase with increase in excess nitrogen. In addition, leaching greater than 30 - 40 kg ha^{-1} will result in nitrate concentrations larger than the EU standard of 50 ppm in the ground water. In the example given (Fig. 4) that would have occurred between application rates of ca 250 and 300 kg ha^{-1}, if there were no contribution from nitrogen in the tops, and at lower levels of

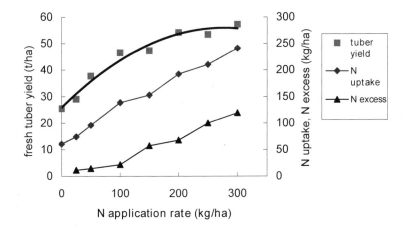

Figure 4   *An example of the response of potato to nitrogen fertilizer application. Left Y-axis:*
           fresh tuber yield (t ha⁻¹). Right Y-axis the uptake of nitrogen in the total crop (tubers
           plus haulm) at harvest (kg ha⁻¹) and the 'N excess' which is defined in the text).
           Data after Vos (1997).

application otherwise. However, it is not possible to establish generally
applicable limits for N excess above which environmental standards are
violated, because the quantity and type of emission (i.e. nitrate leaching or
denitrification) depend on local soil factors and weather.
In general it can be stated that if potato growers are to respect the European
policy on nitrate leaching, they cannot apply nitrogen to the point where the
maximum yield is attained. They have to settle for less. However,
considerations of quality should provide incentives to do this.
There is a weak relation between nitrogen supply and the quality of potatoes.
Most quality attributes of potatoes are promoted by an adequate supply but
levels that are too high impair some. Qualities that are reduced by high N
application are the crisping and frying quality, and storability. The dry matter
concentration also is slightly reduced with high N levels (Jenkins & Nelson,
1992; Vos, 1997). A large supply of nitrogen, leading to small, late increases in
yield also results in a crop canopy that is still growing at the time of harvest.
That is a serious disadvantage for the grower, as well as risking further leaching
of nitrogen from the crop residues.

Plant processes affected by water supply

The expansion of leaves and stems is brought about by water pressure in plant cells. Cells are 'inflated' with water. Reduction in the expansion of stems and leaves is the first symptom of water limitation (Jefferies, 1989), followed by stomatal closure and, much later, reduction in the rate of dry matter accumulation per unit of solar radiation intercepted. That is, the main effect of water restriction on growth is that a smaller leaf canopy intercepts less light. Leaves start to wilt from the bottom of the plant upward. During prolonged drought, leaves also die in the same sequence and more leaves die the longer the period of drought. Potato can survive for some time in a wilted state, except under high solar radiation and high temperature, when the temperature of plant tissues may exceed lethal levels, causing death. Upon re-watering, wilted plants can resume the normal functions quickly. If there are sufficient leaves left to absorb all solar radiation, the crop growth rate will return to the values of non-stressed plants. The growth that should have been made during the period of stress can never be regained.

Sufficient water for unrestricted growth facilitates the crop to use the other resources at their maximum level of efficiency, including nutrients and all the energy spent to manage the crop. For instance, if a crop has a shortfall in yield by 30 per cent because of drought, then the margin of profit over the fixed costs of invested capital, energy and fertilizer will be reduced by much more than 30 per cent and could be eliminated altogether. Moreover, the amount of residual nitrogen must increase and that is a threat to the environment. These aspects need to be considered in policy making for restrictions on irrigation.

Periods of drought are indirectly associated with the phenomenon of second growth. When a temporary stress causes a check in the growth of potato tubers then the control over the seemingly dormant buds on the tuber is lost and these buds may start to grow and form either 'stress sprouts' or a second, junior crop of tubers. Both drought and high temperature (about 28°C and above) can cause second growth. If both stresses are coupled then the effect is aggravated. Preventing water-stress is important, as a drought may break dormancy before the crop is harvested. Then second growth and growth cracking may both occur. Dry soil at around the time of tuber initiation is known to encourage the development of Common Scab (caused by *Streptomyces scabies*) on the tubers. Proper irrigation throughout tuber initiation and early tuber growth is the only effective measure to limit the disease.

Excess of water, too, results in adverse effects. The first is that the potato does not tolerate water logging and anaerobiosis and then the return to aerobic conditions. Root function is quickly impaired, resulting in long-lasting negative effects on the growth rate of the crop. A second point is that over-wet soil can encourage the development of Powdery Scab (caused by *Spongospora subterranea*. Thirdly, as excess water drains through the soil nitrate can be leached below the rooted zone, or denitrified in pockets of soil where anaerobiosis develops. Both processes, leaching and denitrification, may result in N deficiency and certainly cause pollution.

Transformations and losses of nitrogen and processes that affect the supply of nitrogen to the plant

All nitrogen that is present in living and dead organic material, i.e. is present in the biosphere, is drawn from the atmosphere and will eventually return from the biosphere to the atmosphere. Some bacteria can incorporate atmospheric nitrogen into organic compounds. The process is called biological nitrogen fixation. *Rhizobium* bacteria that live in symbiosis with legumes and eventually make nitrogen available to the host plant are a well-known example. Some other soil organisms have a similar capability, but their role is negligible in arable farming. Atmospheric nitrogen can be oxidised by lightning and in combustion. Nitrogenous compounds in the atmosphere generally return to the soil surface in rain or snow, or by dry deposition. Ultimately, the deposited nitrogen becomes available to plants. Figure 5 gives a schematic illustration of the global nitrogen cycle and indicates the relative magnitudes of the pools and fluxes. A more detailed diagram of the nitrogen cycle in the soil in given in Chapter 3.1.

Nitrogen in any part of the biosphere is continually recycled: plant products, whether or not first eaten by animals or humans, return to the soil as organic material. In the soil, dead organic material is degraded by soil micro-fauna and by micro-organisms. Those micro-organisms use some of the nitrogen directly for their own growth and the rest is 'mineralised' into ammonium and, subsequently, to nitrate. Mineralized N is available again to plants. Even the fraction taken up by the micro-organisms is eventually mineralized. Some

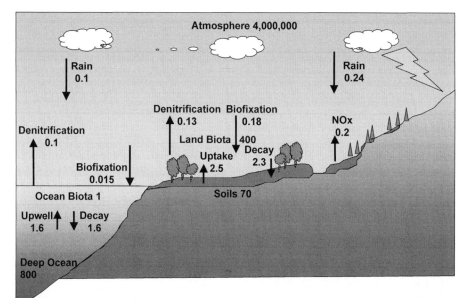

The principal reservoirs of the nitrogen cycle in 10^{12} kg. Fluxes represent the annual rate of transfer.

Figure 5   Schematic illustration of the global nitrogen cycle.

components of the organic matter (e.g. lignins) persist in the soil and degrade at a very slow rate. Soil humus, more or less equivalent to what we call the 'organic matter content' of the soil, mostly comprises material that degrades slowly. Each percentage point of soil organic matter in the top 25 cm of soil represents a pool of about 1700 - 1800 kg N ha^{-1}. However, only two to three per cent of that pool is mineralized annually. Some components of freshly incorporated plant material, e.g. proteins, degrade quickly, that is in weeks or months.

Ammonium is a positively charged ion, which is attracted by the negatively charged surfaces of soil organic matter and clay minerals. Nitrate is a negatively charged ion and is repelled by organic matter and clay minerals. For that reason nitrate moves easily with soil water. When rainfall and irrigation bring the soil water level above 'field capacity', water starts to drain down the soil profile to depth, taking nitrate with it: nitrate is leached. Leaching represents a loss of nitrogen from the system and is an environmental concern. Nitrate rich water upsets aquatic ecosystems, and nitrate-rich groundwater is unacceptable as drinking water.

In the soil, ammonium is usually quickly transformed to nitrate by bacteria; this process is called nitrification. The bacteria are less active at lower temperatures and so the ratio of ammonium: nitrate is higher in winter and early spring than in summer. Nitrate is the principal form of nitrogen taken up by plants but it is also the most mobile form and can be lost from the soil, by leaching or by denitrification into a gaseous form. Nitrification inhibitors have found some application in practice. These substances inhibit the bacteria that transform ammonium to nitrate and so spread the risk of losing nitrate. To estimate future mineralization in soil one needs to know the amount of organic matter in the soil, any fresh organic material that is added, and their nitrogen concentrations. One would also need to know soil temperatures and moisture content.

The micro-organisms in the soil require nitrogen for their metabolism, too, and so they will absorb mineral nitrogen from the soil if there is insufficient in the dead organic material, e.g. plant residues, that they are breaking down. That mineral nitrogen is then said to be immobilized because higher plants, e.g. crops, cannot use it until it is re-mineralized when the micro-organism itself dies. The balance between mineralization and immobilization depends on the nitrogen concentration in the plant residues, expressed as the carbon to nitrogen ratio, C:N ratio. Generally, where the C:N ratio is less than 20, there is net mineralization. Where C:N ratios are greater than 30 there is net immobilization. Intermediate ratios have little or no net effect on the level of nitrate in the soil.

Nitrogen can be lost from the soil-plant system in various ways, including volatilization, denitrification and leaching. Ammonium ions are readily transformed to ammonia, which can escape from the soil. In arable production ammonia losses are generally small, except under particular conditions such as soil pH of 7 or higher. It would be wrong to use ammonium fertilizers on such soils. Spreading manures and slurries on the soil surface also results in large losses of ammonia. So manures and slurries should be placed directly into the soil, or incorporated very rapidly.

There are, also, soil bacteria that can and do reduce nitrate to nitrous oxide (N_2O) and molecular nitrogen (N_2), using the oxygen in nitrate instead of oxygen in the soil atmosphere. This is called denitrification, and results in the loss of nitrogen to the atmosphere. Nitrous oxide is an atmospheric pollutant contributing to global change. The conditions that encourage denitrification include poor drainage, warm soil, neutral soils, and a good supply of readily decomposable organic matter, e.g. fresh crop residues.

The nitrogen budget of the crop

Since Von Liebig discovered in the last century that plants need mineral ions to grow, man has made fertilizers. Nitrogen fertilizers are made from atmospheric nitrogen and have become the dominant source of nitrogen for crops. However, the losses of N from agriculture and the associated undesired ecological effects call for new and improved strategies of nitrogen management in crop production.

The various transformation and transport processes can be grouped into two categories: sources and sinks of nitrogen. Sources include the mineral nitrogen initially available, nitrogen becoming available through processes in the soil and any external supply such as fertilizer. Sinks diminish the amount of mineral nitrogen in the soil. Mineral nitrogen left at harvest is called residual nitrogen and is regarded as a sink term as well. Table 1 presents an overview of sinks and sources.

Proper nitrogen management entails quantifying each of the items on the budget so that losses and residual nitrogen may be minimized, while the use of nitrogen from all sources is maximised.

Sources of water for the plant

There are three possible sources of water for the crop: rainfall, irrigation, and the water stored in the soil. During the growing season, daily evapotranspiration is not replenished daily. Instead the crop uses water stored in the soil. The amount of water stored and available to the crop is limited by the maximum and minimum soil moisture contents. These limits are strongly determined by soil texture (e.g. sandy soils hold least, loamy soils are intermediate, and clay soils hold the most in a given depth). Above the upper limit of the water content of the soil it starts to drain. The lower limit is the water content at which plants start to wilt. The depth of soil that is explored by the roots also determines how much stored soil water is available for the crop. On soils with high groundwater tables, capillary rise may increase the amount of soil water that is available to the crop.

Table 1   *The nitrogen budget of the soil. Sinks add nitrogen to the pool of mineral nitrogen in the soil; sinks remove mineral nitrogen from the soil (further explanations in the text).*

Sources of N	Sinks of N
(i) Initial mineral N at planting	(vi) Uptake by the crop
(ii) Net mineralization	(vii) Denitrification
(iii) Atmospheric deposition	(viii) Volatilization
(iv) N in organic manures	(ix) Leaching
(v) N in fertilizers	(x) Residual mineral N

The following comments apply to the items on the nitrogen budget in Table 1:

(i)   The initial amount of mineral nitrogen (= ammonium + nitrate) in the soil at planting is measured by soil analysis; laboratories specify different depth for sampling soil nitrogen, e.g. from 0 - 30 cm to 0 - 100 cm depth.

(ii)   The amount of net mineralization during the growing season is the difference between 'gross' (real) mineralization and immobilization. When making calculations it is relevant to distinguish nitrogen mineralization from the large bulk of soil organic matter (high C:N ratio) and mineralization of freshly incorporated manures or plant debris (low C:N ratio).

(iii)   Atmospheric deposition depends on the intensity of production, i.e. traffic and animal production. In most European countries the annual rate of deposition is close to or exceeds 30 kg N ha^{-1} y^{-1} (Schleef and Kleinhanss, 1996). Atmospheric deposition during autumn and winter cannot be used by the crop.

(iv)   Nitrogen is incorporated with manures in different forms: as readily available nitrate and ammonium, and as N in organic substances.

(v)   Ammonium and nitrate are applied as chemical fertilizers.

(vi)   Uptake by the plant is self evident; it is important to realise that about 20 to 25 per cent of the nitrogen absorbed by the crop remains on the field as crop residues.

(vii) Denitrification, volatilization and leaching are losses; these processes need to be minimized.

(x)   The amount of residual mineral nitrogen should be minimized. Residual nitrogen is not directly lost, but in most European countries residual nitrogen is not retained in the soil until the next spring but is lost through denitrification or leaching. Potatoes are known to leaving comparatively high levels of residual N. It may be possible, in principle, to grow nitrogen catch crops after potato. However, the later the date of potato harvest and the higher the latitude of the location, the less the scope for growing a catch crop.

Plants suffer from drought when the cumulative amount of rainfall in a particular period of time plus the amount of available soil water are less than is required for evapotranspiration in that period. Indeed, they will suffer before that amount of water is used up. For any given environment, the larger the soil storage capacity the lower the chances of water-stress in the crop. Figure 6 shows long-term averages of rainfall and potential evapotranspiration for 10-day periods throughout the year in The Netherlands as an example. The graph shows that, on average, rainfall exceeds potential evapotranspiration from September through March. From April through August, on average, a green crop needs more water than is available from rain. The average deficit is 80 mm. It would seem that, if the amount of soil water available to the crops is 80 mm or more then, in an average growing season, water-stress would not occur. Reality, however, is more complicated than that. Not all the 'available' soil water is immediately accessible, the roots must grow through the soil. Further, 'average' totals do not indicate differences in timing of supply, yet the losses in crop yield and quality also depend on the timing of a stress period.

Proper water management entails keeping a soil water budget during the growing season. The relevant items in the budget were shown in Figure 1.

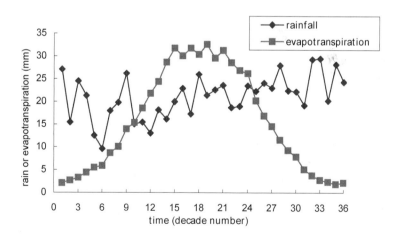

Figure 6   *Average values (mm per decade of days) of rainfall (full drawn lines and squares) and potential evapotranspiration (broken lines and circles) in The Netherlands for 1961-1990. The numbers 1, 2, 3, ..., 33, 36 represent values for days 1-10, days 11-20 and so on through the year, respectively.*

The gains are self-explanatory; the losses are processes removing water from the rooted zone. Runoff is the flow of water over the surface of sloping fields and should be minimized. The diagram makes evident that drainage occurs whenever rainfall or mis-applied irrigation eliminate the soil moisture deficit and the water content of the soil is above field capacity. Likewise drought and water-stress occur when the losses exceed the gains by most of what is stored in the soil.

Principles of nitrogen management: dealing with the unknown

The grower's objective is to provide sufficient nitrogen to achieve the potential yield of the crop, given the environmental conditions and the crop's genotypic properties. Basically, the problem of the grower is to assign values to the items of the nitrogen budget of Table 1, and to items (iv) and (v) in particular: N to be added with organic manures and fertilizers. The nitrogen requirement of the crop is associated with the yield but the grower does not know the yield in advance, and so cannot know exactly the nitrogen requirement. Nor is it known in advance how much nitrogen the soil will supply. Yet excess nitrogen is at risk of being lost to the environment where it would be a pollutant.

With a single nitrogen application at the beginning of the growing season, it is virtually impossible to get right the calculation of the seasonal N requirement because the yield, and perhaps also the nitrogen concentration in the crop, depend on the weather and on biotic and abiotic constraints in the field. The rate of net mineralization depends on the C:N ratio of active organic pools and on soil temperature and soil moisture which vary between and within seasons. The efficiency with which nitrogen is used is variable since it is affected by events such as immobilization, leaching and denitrification in anaerobic pockets of soil.

Splitting the nitrogen application offers the prospect of getting the calculation right. This is because, as the season progresses, one can make closer and closer estimates of final yield. An estimate of yield provides an estimate of the nitrogen requirement of the crop, although the average nitrogen concentration appears to show substantial variation across years (Vos, 1996). With split applications the grower can divide the growing period into smaller periods of time. At the beginning of each period a decision is taken on how much nitrogen is required till the next point of decision or until the final harvest. The grower needs to ensure that sufficient nitrogen is available during each of these periods. That means that for each period values need to be estimated for each of the items on the N budget (Table 1).

At each point of decision the grower can make the calculations only if there are direct or indirect estimates of N uptake and crop mass. Likewise, an estimate is required of yield and the nitrogen content and mass of the crop at the next point or at final harvest. Analysis or estimates of changes with time in soil mineral nitrogen may be needed as well. There are a number of important requirements for methods to estimate the current nitrogen status of the crop and the soil. These include:

- acceptable labour requirement
- economic in price
- short period of time between start and completion of the test.

A number of those analytical methods are described and evaluated in later chapters of this book.

Principles of water management

In areas where there is little or no rainfall during the growing season, irrigation is necessary if potatoes are to be grown. Elsewhere, if there is a source of water available - river, borehole, reservoir - the potato grower should make the strategic decision whether or not to invest in irrigation equipment. This decision will be based on the frequency and extent of periods without rain, evaporation rates during dry weather, and the water holding capacity of the soil. The balance between these three indicates the likelihood of the potato crop being water-stressed. Other factors to be considered include the capital cost of the equipment and its installation, the cost of extracting water, the cost of an extraction licence (if applicable), the cost of labour, and the likely gain in yield and quality of the crop. This is not the place to detail the calculations that must be made to make that decision. The factors are considered more fully elsewhere in this manual, and there are computer programmes available to help the grower come to a decision. It is worth noting, however, that the type and standard of soil tillage practised can affect the amount of water held in the soil. This is explained in Chapter 5.

Here, we want to reinforce the principle that the potato crop grows best, yields the highest, and produces the best quality tubers where the supply of water to it can be controlled. Further, an adequate water supply is needed if the crop is to use the soil nutrients effectively. In most potato growing areas, these aims can only be achieved if irrigation is available. If it is not available, for whatever reason, then the potato grower has to accept that his or her enterprise is exposed to much more risk and uncertainty.

The second set of principles concerns the operational decisions of whether to apply irrigation on any given day and, if so, how much water to give. Where there is a large area planted to potatoes and limited capacity for irrigation the grower must decide also on the sequence in which fields are to be irrigated and, even, on whether to omit some from the irrigation cycle.

Water management at the operational level is based on maintaining a daily water balance or budget for each field (Fig. 1). Losses of water through evapotranspiration (estimated from meteorological data) and gains through rainfall and irrigation are tallied and the need for irrigation is recognised once when they move out of balance by a certain amount, the soil moisture deficit. Two points should be recognised here. The first is that rainfall, particularly in summer, is very variable from place-to-place. Showers from cumulo-nimbus clouds, for example, are highly localised. Therefore, a grower should record the rainfall in each individual field that is to be irrigated. The second point is that the 'threshold' soil moisture deficit to trigger irrigation will differ from field-to-field with the water holding capacity of the different soils, and it will change continuously through the growing season as the crop develops. If the soil and tillage allow roots to continue growing downward then there is a larger reservoir of soil water made available. In addition, the crop becomes less sensitive to water stress as the growing season progresses and so larger soil moisture deficits can be tolerated. This ability of the crop should be exploited by allowing it to use up a large part of the available soil water by the end of the season. The reservoir of soil water is a resource that is there to be used and a grower should not irrigate to maintain a moist soil through to harvest.

For irrigation systems that are transported from field to field, or where there is a limit on what area can be irrigated in a day or how much water is available in a day or a week, questions arise on how best to manage irrigation scheduling. These questions can only be answered through the use of computer programmes. These issues and those considered in the preceding paragraphs are presented more fully in Chapters 4 and 5 of this manual.

Here we have only considered the water budget of the crop. However, there are instruments and sensors available that will measure soil water or plant water status. These can be used to indicate impending water shortage and water-stress, respectively. These and other methods are treated and evaluated in Chapter 4.

Conclusions

An optimal nitrogen supply ensures unrestricted growth and a yield of tubers that is close to potential, while minimising losses to the environment. However, calculating that optimal rate of nutrition is not straightforward, and it cannot be done at or before planting. This is because the supply of nitrogen from the soil is modified by several processes of transformation, transport, and emission that are dependent on the actual weather experienced, among other things. Splitting the supply of nitrogen over two or more occasions greatly increases the opportunity to match the total supply to the actual nitrogen requirement for fertilizer. If the opportunity is to be fully realised, then the grower must monitor the nitrogen status of the soil and the crop during the growing season. Methods to do this are discussed in the succeeding chapters of this manual. The quality of the tubers produced is at its best where the supply of nitrogen has just matched the requirement of the crop over the length of growing season. This is broadly indicated by the onset of canopy senescence shortly before the planned harvest or date of haulm destruction. Therefore, we recommend managing the nitrogen nutrition of the crop so that the haulm is in decline at close to harvest and so that haulm destruction is not a major task.

Decisions on irrigation (and soil improvement) need to be taken at two levels: (i) the strategic level of investment and (ii) the operational level of using the irrigation system at a given time. The strategic decision depends on probability of drought and water-stress (determined by probabilities on rainfall and the water holding capacity of the soil), and the economics of losses in yield and quality versus the fixed and variable cost of operating the equipment. The operational decisions on irrigation should be guided by a rational system of calculation or measurement. It should not be based on 'hunch' or the feel of the soil. An adequate supply of water is needed for the more efficient use of other resources, including nitrogen.

2 Plant nitrogen status

2.1 The significance of trends in concentrations of total nitrogen and nitrogenous compounds

G. Gianquinto & S. Bona

When optimizing nitrogen nutrition of the potato crop, it is of prime importance to consider plant nitrogen concentration throughout the growing season. Relevant basic knowledge must first be described to answer basic questions such as: What is the total requirement of the crop for nitrogen? What is an expected trend of nitrogen concentration in the different organs during the growing season? How is this trend modified by soil characteristics and climate? Do cultivar characteristics and technical practices modify the trend?
Then to be useful in decision making in fertilization strategy, specific other questions must be answered: When is plant nitrogen status considered as sufficient? What are critical values or ranges for plant nitrogen? What is the plant reaction time to nitrogen deficiency? What type of plant analysis is the best for making fertilizer recommendations? What is the plant's reaction time to supplementary nitrogen? How optimize and standardize the methodology of analysis? This chapter answers all these topics as a general introduction to the use of plant nitrogen concentration assessment.

Introduction

It is known that the maximum yield of high quality potatoes requires adequate nitrogen nutrition throughout the growing season (see Chapter 1.2). Both low and unnecessarily high rates of nitrogen fertilizer can adversely affect the crop, e.g. reducing growth and yield or delaying tuber fill and maturity, lowering tuber specific gravity, and compromising storability. The use of high rates of fertilization may also have an adverse effect on the environment, increasing the potential rate of nitrate leaching. So, the estimation of the requirement of the crop for nitrogen becomes very urgent both for economical and environmental reasons.

Rules of thumb were used for a long time as the principal method for many farmers and agricultural advisors to assess the amount of nitrogen fertilizer to apply. This method cannot give the best answer in most cases. Other methods have been developed to provide a correct approach in the assessment of crop nitrogen status and to improve the management of nitrogen input. These methods can be based on direct or indirect measurements of soil and/or plant nutrient concentration. Soil tests, carried out before planting, assess the nitrogen already available for the crop and provide an estimate of the additional amount of nitrogen to apply. Plant measurements are more suitable for monitoring the crop's nitrogen requirement throughout the growing season, and allowing adjustment for errors in fertilization. They can be used to assess if the initial supply was sufficient, thus indicating if and when the crop needs supplementary nitrogen. The problem of these measurements is whether current soil or plant nitrogen concentrations can indicate whether the supply will be adequate in the future. Soil tests are discussed in Chapter 3, and the methods to determine plant nitrogen status are presented in Chapter 2.2 (invasive) and Chapter 2.3 (non–invasive). Some determine directly the concentration of organic (Kjeldahl), inorganic (e.g. Petiole Sap Analysis), protein- (e.g. Analysis of Rubisco) or total nitrogen (micro Dumas). Other methods can be used to deduce the nitrogen concentration from measurements of ground cover or leaf reflectance in different wavelengths (chlorophyll-meter, Near-Infrared Reflectance). Growers can use some of these methods but others are more labour intensive and expensive and only a few laboratories have the necessary equipment. Depending on analytical procedures these measurements can provide nitrogen concentration per unit of either fresh or dry weight. Basically, nitrogen concentration expressed per unit of dry weight is more reliable – even though the functioning nitrogen is in solution – since it is independent of plant water status.

Whatever the method chosen, the best application of plant analysis depends upon the detection of a Critical Value or Range. Above the Critical Value or Range crops are not expected to respond to the addition of nitrogen. Below the Critical Value or Range the nitrogen requirement should be closely related to the deviation from it. However, establishing an optimum nitrogen concentration or Critical Value or Range is quite difficult because nitrogen concentration varies with plant part and with stage of development. Moreover, optimum nitrogen concentration may vary with genotype and climatic or technical factors, other than the nitrogen supply.

The basic objective of this chapter is to provide a description of trends of nitrogen concentration in plants during growth, and subsequently to examine what variations in nitrogen concentration would be expected by cultivating the crop in different environmental and/or technical conditions, attempting to provide general principles which can help in decision making.

Trends of concentration of nitrogenous compounds in plant organs during growth

Before starting to evaluate the effect of different environmental and/or agronomic factors, the role of the most relevant nitrogen compounds needs to be considered as well as their expected trends in different plant organs during the growing season. Nitrate is the major source of inorganic nitrogen taken up by the roots. It is readily mobile in the plant transport system (xylem) and can also be stored in cell structures (vacuoles) of roots, shoots, and tubers. The nitrate accumulated in vacuoles is important for various processes (e.g. exchange and balance of nutrients, and osmotic adjustment) and for the quality of tubers as, at high concentration, nitrate is toxic to humans and animals. After reduction to ammonia, most of the nitrate is incorporated into organic compounds. Organic nitrogen compounds can be classified into two main categories: low molecular weight (amino acids, amides, peptides, amines, polyamines, etc.) and high molecular weight compounds (proteins, nucleic acids, enzymes, etc.). The first compounds are not only the constituents of the second ones, but also serve many functions in plants (e.g., long distance transport of reduced nitrogen, secondary messengers, and membrane protection). Proteins and other high-molecular-weight compounds are involved in almost all the processes within plant cells, so they are basic for potato growth and production.

Due to the importance of nitrogen compounds in plant function, an overview of their concentration in plant tissues is needed. Table 1 shows concentration trends throughout the growing season of the most important or more easily detectable nitrogen compounds in potato organs. Experimental data, including different cultivars, years and locations were collected (Kolbe & Stephan-Beckmann, 1997; Osaki et al., 1993.), adapted and analysed to represent general trends.

Table 1   *Development over time (DAE=days after emergence) of fresh (FW), dry weight*
             (DW), and percentage of dry matter (DM), total, organic, and inorganic nitrogen in
             potato plant organs. Beginning of flowering = 40–45 DAE; 50% maximum tuber
             growth = 70–75 DAE; beginning of senescence = 100–105 DAE. (Adapted from
             Kolbe & Stephan-Beckmann (1997), and Osaki (1993).

DAE	FW	DW	DM	Total N	Organic N	Protein N	Rubisco N	Non Protein N	Nitrate N
	(g)	(g)	(%)	(% DM)	(% DM)	(% DM)	(% DM)	(% DM)	(% DM)
Leaves									
15	57	5	9.1	7.19	6.15	4.81	1.151	1.34	1.040
45	702	68	9.7	4.69	4.39	3.76	0.780	0.61	0.300
75	520	62	12.0	3.58	3.56	2.80	0.572	0.26	0.019
105	190	42	21.0	2.52	2.50	1.90	0.353	0.29	0.018
120	152	37	24.2	2.23	2.20	1.74	0.201	0.41	0.030
Stems									
15	28	2	6.1	7.09	4.94	3.41	N.D.	1.02	2.153
45	515	44	8.6	2.79	1.95	1.50	N.D.	0.23	0.840
75	486	53	11.0	1.39	1.31	1.09	N.D.	0.08	0.076
105	280	37	13.1	1.27	1.24	0.95	N.D.	0.06	0.032
120	220	33	14.8	1.30	1.26	0.96	N.D.	0.08	0.043
Tubers									
30	48	7	14.6	2.32	2.29	1.29	N.A.	1.00	0.031
45	335	54	16.1	1.81	1.80	1.00	N.A.	0.81	0.013
75	1280	288	22.5	1.55	1.54	0.82	N.A.	0.65	0.004
105	1575	378	24.0	1.48	1.47	0.83	N.A.	0.64	0.004
135	1582	356	22.5	1.59	1.58	0.88	N.A.	0.70	0.006

N.D. = not determined; N.A = not applicable

Leaves

From 15 days after emergence, fresh and dry weights rapidly increase.
Maximum weights can be observed between days 45 and 75, then values
slowly decrease. The highest concentrations of total nitrogen (organic +

inorganic), and several nitrogen compounds, are found in very young plants. Total nitrogen concentration continuously decreases through the growing season, although the greatest reduction occurs between 15 and 75 days after emergence. The time courses for organic and protein-nitrogen are similar to that of total nitrogen, but concentrations decrease more slowly during the first 45–75 days of growth. The trend of Rubisco-nitrogen concentration is quite similar to that of protein-nitrogen. Rubisco (ribulose 1,5-bisphosphate carboxylase/oxygenase) represents a large proportion of leaf proteins and plays an important role in photosynthesis. The concentrations of non-protein organic nitrogen (including other organic compounds, such as free amino acids, amides, etc.) and nitrate-nitrogen decrease continuously and strongly until 75 days after emergence. The concentrations of free amino acids and amides often increase, to some extent, at the end of the season because protein-nitrogen is catabolized and accumulated in the leaves, but not translocated.

Stems

The time courses for fresh and dry weights are similar to those described for leaves. The total nitrogen concentration in the stems of very young plants is almost the same as in leaves but, considering the distribution between compounds differs with higher nitrate-nitrogen and lower organic nitrogen concentrations. Furthermore the concentrations of all nitrogen compounds tend to decrease more rapidly than in leaves during main canopy growth (15–75 days after emergence); thereafter they remain fairly constant.

Tubers

Most of the increase in fresh and dry weights occurs from 30 to 90 days after emergence. Nitrogen concentration declines with advancing maturity, although the changes are smaller than in the shoot. As a rule, concentrations of total, protein-, and nitrate-nitrogen are lower than in leaves and stems (at least during main canopy growth), while the concentration of non-protein organic nitrogen is higher.

The foregoing description shows that analysis of leaves and stems should be considered more appropriate than analysis of tubers analysed for plant nitrogen status. They can be sampled more easily and the magnitude of detectable values during the main period of canopy growth reduces the incidence of measurement errors (due to instrument sensitivity, the approach of the operator, sampling procedure etc.). However, that leaves the questions of which portion of shoot has to be chosen for measurements and how to sample? For relevant answers, refer to Chapter 2.4. The nitrate status of petioles and, in recent years, the chlorophyll concentration of leaf blades (chlorophyll meter measurements - SPAD or HNT), which is correlated to nitrogen concentration, are probably the most intensively studied potato nutritional indicators. Therefore, in the following sections, these indicators will often be used to illustrate the influences of environment and crop management on the nitrogen status of plant.

Variations in trend of nitrogen concentration related to nitrogen supply

Rates and timing of nitrogen application

Examples of the time courses for total nitrogen concentration at different levels of nitrogen application at planting are shown in Figure 1. For both leaves and stems, at the first sampling date (50 days after planting) only the treatment receiving the highest nitrogen rate (300 kg N ha^{-1}) clearly differs from the non-fertilized treatment. Sampling after day 50 allows clear separation of the treatment receiving 100 kg N ha^{-1} from the non-fertilized treatment. Sampling after day 80 provides a good separation between treatments receiving higher rates of fertilizer (200, 300 kg N ha^{-1}). Differences between treatments receiving 100 and 200 kg N ha^{-1} are more evident in the stems.

Similarly, a clear separation of treatments receiving medium or high rates of fertilizer is detectable quite late in the season using concentration of nitrate-nitrogen in petioles, and SPAD values (Porter & Sisson, 1991 and 1993; Vos & Bom, 1993). These results suggest that early sampling may not accurately detect suboptimal nitrogen status in potato plants that have received moderate levels of fertilizer at planting, but may be used to identify extremely nitrogen deficient crops.

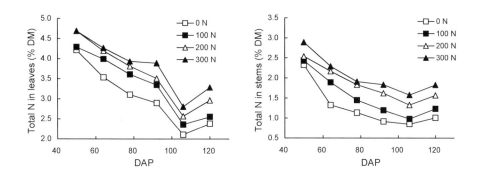

Figure 1   *Total nitrogen concentration in leaves (left) and stem (right) of 'Lutetia' potato, as affected by N rates (kg ha⁻¹). DAP = days after planting. (Guarda and Colauzzi, unpublished data).*

Concentration of nitrate-nitrogen in petioles generally declines during sampling period, but it is either maintained or increases following additional application of nitrogen (Fig. 2): it is often maintained at higher levels than treatments receiving all nitrogen at planting.

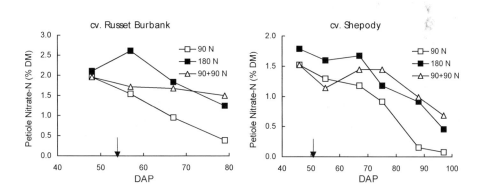

Figure 2   *Petiole nitrate-nitrogen concentration of 'Russet Burbank' and Shepody' potatoes, as influenced by rate (kg ha⁻¹) and timing of nitrogen application. Nitrogen applied at planting (90N and 180N), and split half upon planting and half after tuber initiation (90+90N). Arrows indicate timing of additional application. (Adapted from Porter & Sisson, 1993).*

Nitrogen sources and placement methods

Petiole nitrate-nitrogen concentration in crops fertilized either with ammonium sulphate or ammonium nitrate is constantly higher than that found in a crop fertilized with calcium nitrate (Bundy et al., 1986). Differences found between fertilizer types are consistent with greater nitrogen losses through leaching from the fertilizers containing nitrate. These effects can be enhanced or reduced by conditions of soil moisture. Also the placement method used to apply supplemental nitrogen treatments (band application near the row and incorporation under the shoulder of the potato hill) should minimize nitrate leaching. However, it has been observed that plant nitrogen concentration is not appreciably different whether fertilizer is banded at planting or broadcast before planting (Timm et al., 1983).

Variations in trend of nitrogen concentration related to environment

Variation across sites is to be expected due to differences in nitrogen supplying capability of soil and weather conditions. Many studies have shown that the relations between rate of nitrogen application and tissue nitrogen concentration are not constant across sites and growing seasons.

Soils

In experiments carried out in Italy, the interaction between rate of nitrogen application and soil type is evident (Fig. 3). Differences between soils are particularly clear in non-fertilized or moderately fertilized plots (50 and 100 kg N ha^{-1}). At higher nitrogen doses the influence of soil becomes less relevant. In moderately fertilized plots, SPAD values are higher in sandy soil than in loamy soil during the first period of growth (prior to 76 days after planting), due to the greater amount of nitrogen that is readily available to the crop. Then SPAD values decrease more rapidly in crops on sandy soil, as it is more susceptible to losses, including leaching. At higher nitrogen doses, losses obviously occur but the amount of nitrogen remaining in the soil is sufficient to feed the plants. These effects can be enhanced or reduced by other factors such as rainfall, irrigation and temperature.

Temperatures and light conditions

Temperature and light conditions (solar irradiance and daylength) are two of the main factors influencing plant development, so they can affect tissue nutrient concentration. For example, cold weather that slows potato growth causes increased nitrate concentration in petioles compared to unstressed plants. The role of increased nitrate concentration in tissues of plant grown at low temperatures, in enhancing plant resistance to cold is well known. Conversely, warm weather that accelerates plant growth can cause a dilution of nutrients in the tissues. High temperatures can also reduce nitrogen uptake indirectly, by reducing the amount of nitrogen available in the soil due to reduced availability of water.

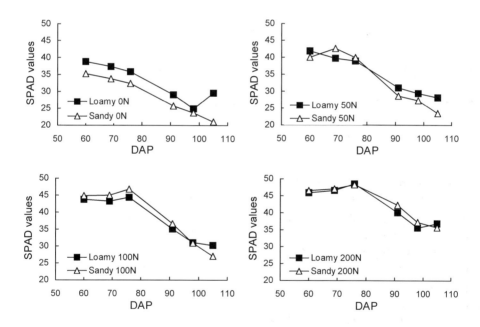

Figure 3   *Development over time of SPAD values in 'Primura' potato grown in differing soils and with several levels of nitrogen supply (kg ha⁻¹). (Gianquinto, unpublished data).*

Variations in trend of nitrogen concentration related to crop management

Cultivars

The level of plant nitrogen can be cultivar dependent, particularly when sampling is based on chronological age. In Figure 2 the influence of two different cultivars and nitrogen fertilization on petiole nitrate-nitrogen concentration was shown. These cultivars respond differently to conditions imposed by rate and timing of nitrogen application and sampling date.

It is commonly accepted that early-maturing cultivars may exhibit lower nutrient levels than later-maturing cultivars sampled on the same date, probably because the former are physiologically more mature (Lewis & Love, 1994). However, experiments conducted in Italy have shown that differences in total nitrogen concentration among varieties belonging to different maturity classes may be less evident than differences among cultivars belonging to the same class of maturity. This is attributable to the Mediterranean climate, characterised by dry and hot summer, which shortens the growing season of later-maturing cultivars, advancing maturity and senescence.

Soil fumigation

Soil fumigation can have a pronounced effect on nitrogen concentration of plants. E.g., petiole nitrate-nitrogen concentration from non-fumigated plots can be about 15% higher than levels in petioles from plots fumigated with chloropicrin (O'Sullivan & Reyes, 1980). This may be explained as the effect of the chemical killing the nitrifying bacteria so that soil nitrogen persists in the ammonium form.

Crop rotation

The rate of decline in plant nitrogen concentration may vary in time, largely depending on the previous crop. Where a leguminous crop precedes potatoes, the concentration of nitrate nitrogen in petioles may decline much more slowly over the time than after a winter cereal (Fig. 4). This is attributable to relatively slow mineralization of the nitrogen-rich legume residue. Moreover, during

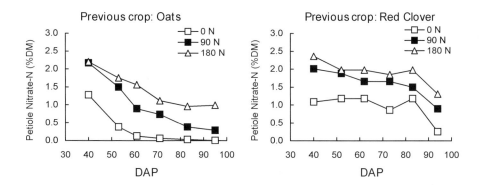

Figure 4   *Petiole nitrate-nitrogen concentration over time in 'Russet Burbank' potato, as influenced by nitrogen rates (kg ha⁻¹) and previous crop. (Adapted from Porter & Sisson, 1991).*

straw decomposition, the high C:N ratio of cereal residue can cause immobilization of nitrogen and decrease nitrogen availability for the following crops. However, in light soil, this negative effect of straw residue may be balanced by an improvement of some physical characteristics, such as soil water capacity.

Tillage

In heavy soils stratification due to compaction by tillage implements frequently occurs. Soil compaction must be carefully avoided, because it retards the growth of potato roots and restricts their development to the upper 25–30 cm of soil. Deep tillage, disrupting a stratified layer and increasing rooting depth may improve nitrogen uptake. The results of several years experiments have shown that levels of nitrogen in petioles may be increased by a combination of deep tillage and deep fertilizer placement.

Planting date and plant spacing

When the minimum requirement for nitrogen is assured, the planting date should not influence nitrogen concentrations in leaf and petiole. However, when nitrogen is insufficient, a lower concentration might be expected following early planting, as yield response to nitrogen is generally greater after early plantings (longer growing period) (White & Sanderson, 1983). The plant spacing slightly influences nitrogen concentrations in leaf and petiole tissue (slightly lower concentrations are generally observed in the closest spacing).

Irrigation

In plants grown under deficient water conditions the petiole nitrate-nitrogen is usually more concentrated than in unstressed plants. The petiole nitrate-nitrogen concentration may increase by 20-30% or more compared to a well-watered crop, when irrigation replaces only 30-40% of evapotranspiration (ET) (Stark et al., 1993). Conversely, irrigation (or rainfall) exceeding ET usually causes a strong reduction of nitrate-nitrogen concentration. Irrigation treatments giving 1.2 and 1.4 times ET have been found to decrease concentrations by about 10 to 30%, compared to the 1.0 ET treatment. This is mainly attributable to the effect of excess irrigation restricting the root zone – in which available soil

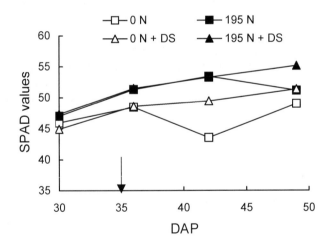

Figure 5   SPAD values over time in 'Vivaldi' potato, as affected by nitrogen rates (kg ha^{-1}) and drought stress (DS). Irrigation restoring 100% transpired water in control, and 50% in DS plants. Arrow indicates beginning of drought stress. (Gianquinto, unpublished data).

nitrogen is rapidly depleted – and in increasing leaching of soil nitrate. Therefore, excessive water supply increases the amount of nitrogen fertilizer required to maintain adequate concentration of nitrate-nitrogen in the petiole.

Figure 5 shows SPAD measurements (see Chapter 2.3) of stressed and well-watered plants at different rates of nitrogen application. Higher values can be observed in the stressed plants. Differences between stressed and unstressed plants are particularly evident in the non-fertilized plots. Moreover, during the stress treatment, SPAD values increased so much that, from 42 days after planting, the unfertilized plants give similar or higher values than fertilized and well-watered plants. At low water availability, plants reduce growth (see chapter 1.2), therefore a low nitrogen availability could be sufficient to sustain this reduced growth (Dalla Costa et al., 1997). Higher SPAD values in drought affected crops are in agreement with practical observation: crops suffering from drought show a darker green colour.

Potassium and phosphorus fertilization

Some experiments suggest that nitrogen uptake can be influenced by potassium and phosphorus fertilization, but information on the response of plant nitrogen concentration to potassium and phosphorus supply are very scarce. However, it has been demonstrated that increasing rates of potassium chloride fertilizer sharply decrease nitrate-nitrogen petiole levels, but fertilization with potassium sulphate has no effect on it (James et al.,1994). The effect of potassium chloride may be explained by the mutual antagonism between NO_3^- and Cl^- during absorption by roots.

Purpose of measurement of nitrogen concentrations in plant organs

After reviewing the changes in nitrogen concentration with time of the potato crop several purposes for measurements can be identified: yield forecast (quantity and quality) and fertilization advice.

Yield forecast

It is evident that yield forecast is crucial in crop management, since farmers can steer their choices in some technical practices (fertilization included) into the light of the expected results. Many experiments have demonstrated that at certain stages of crop development the nitrogen concentration in plant tissues, as well as the concentration of other nutrients, may be correlated with quantity of tubers harvested (Kolbe et al., 1990). In the case of nitrate-nitrogen concentration in petioles and midribs sampled during main canopy growth, positive relations have been found, whereas for leaflets the relationship was less clear. Information collected during experiments may be used by researchers to develop models that, if used properly, permit the prediction of the final yield of the potato crop by farmers or advisors.

Variations of graded yields and N compounds, as well as ascorbic acid, sugars and cell wall components proved to be related to nitrogen concentration in harvested tubers (Kolbe et al., 1990). It should be of great interest to have models drawing relations between these parameters and nutrient concentration during the early stages of tuber growth, in order to estimate final tuber quality and to adjust nutrient input to satisfy special quality demands, e.g. for industrial processing of potato tubers.

Advice on the need for supplemental nitrogen-dressing

Ideally, the grower must match nitrogen availability to the crop's need at each growth stage. Traditionally, all nitrogen is applied preplanting or at the time of planting. However, in recent years many growers apply a portion of the nitrogen requirement during crop growth, applied side- or top-dressed or through irrigation water. Alternatively, nitrogen may be applied via a foliar spray although, because of the low concentrations required to avoid damage, this may require several applications. If additional nitrogen fertilizer is given in solid form, it must be applied in such a way as to reach a moist layer of soil where active roots are present. This may be impossible if the soil is dry, or the crop is too advanced to allow nutrients to be incorporated by cultivation. A low application preplanting with supplemental nitrogen given during the growing season has the potential to reduce losses through leaching, to increase fertilizer efficiencies and to promote early tuber growth. The correct assessment of the nitrogen required by the crop during growth depends upon

Table 2   *Indicative levels of adequate nitrogen in petioles and leaf blades (% DM). (Adapted*
             from Walworth & Muniz, 1993).

Period	Nitrate nitrogen in petioles	Total nitrogen in petioles	Total nitrogen in leaf blades
Early-season (until tuber set)	0.94 - 2.50	3.50 - 7.00	6.00
Mid-season (from tuber set to 50% tuber growth)	0.53 - 3.17	1.42 - 6.00	5.00
Late-season (from 50% to 100% tuber growth)	0.35 - 1.80	2.25 - 5.00	4.00

the availability of Critical Values or ranges, below which crop is responsive to additional fertilization – and a nitrogen requirement closely related to the deviation from the critical level – and above which yield is not affected or may be significantly reduced by addition of nitrogen. The plant nitrogen status can be considered sufficient when the plant has a normal appearance and adequate nutrient concentration for maximum yield. Data from research reports (Walworth & Muniz, 1993) and plant analysis manuals indicate levels of sufficiency in nitrate-nitrogen and total nitrogen in petioles (Table 2).

It can be seen that the variability of the ranges is very large, due to differences in cultivar, crop management, environmental conditions, physiological age of the plant at the time of sampling, etc. It is evident that farmers will need reliable models calibrated for each cultivar, environmental condition and crop management that can be adopted in specific areas.

Figure 6 shows two ways that can be used for representing critical values or range, from which deficient values at 58 days after planting are those below 1180 mg l^{-1} sap nitrate-nitrogen (Waterer, 1997) and 43 SPAD.

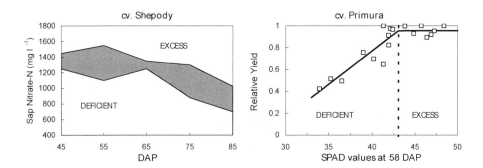

Figure 6   *Critical values of nitrate-nitrogen concentration in petiole sap (cv. Shepody) and*
             SPAD (cv. Primura) with time after planting. (Waterer, 1997, and Gianquinto,
             unpublished data).

Plant nitrogen status and fertilizer requirements (plant reaction time to nitrogen deficiency and to supplementary addition)

Plant reaction time to nitrogen deficiency

Throughout the period of emergence, the plant is dependent on the reserves in the seed–tuber. After emergence, leaf expansion is rapid and the plant quickly becomes autotrophic, so its sensitivity to environmental factors and the supplies of water and nutrients increases. However the transfer of reserves from the seed-tuber continues until it is almost completely depleted or until it decays. Hence, even if available nitrogen in the soil is low, there is no abrupt deficiency of nitrogen supply before 2–3 weeks after emergence. Later, nitrogen deficiency may become more or less evident depending on soil characteristics (Fig. 3), climatic conditions and crop management. In temperate zones and in soils with sufficient mineralization capacity, the potato plant does not display nitrogen deficiency abruptly, and only very dry periods bring plant physiological activity to a stop where there is no irrigation. As a consequence, the detection of early deficient nitrogen conditions in potato crop requires precise methods of analysis of plant nitrogen status.

Plant analysis for correction of nutritional problems

The methods for plant analysis have made it possible to determine nitrogen concentration precisely, quickly and cheaply in the field to make fertilizer recommendations. The strict sampling procedures required and the time lag between sampling and obtaining results may limit the practical usefulness of the procedure. Early season samples are obviously most useful to correct a nutrient deficiency in the current season. Moreover, samples must be carefully taken as the position of the plant organ – whatever plant part is used for analysis – can influence nutritional diagnosis. Figure 7 shows the variation in sap nitrate concentration in petioles with leaf position; concentration increases with depth into the canopy in leaves of both main stems and side branches. With some methods (e.g. analysis of concentration of nitrate-nitrogen in petiole), the time of the day when samples are taken also becomes important. A few experiments describe the diurnal variation in nutrient concentration, but a recent one (MacKerron et al., 1995) reports that the most stable period for concentration in petiole sap appears to be between noon and 3 p.m., at least if high rates of basal fertilization have been used. At moderate nitrogen rates the changes are less obvious both before and after that period.

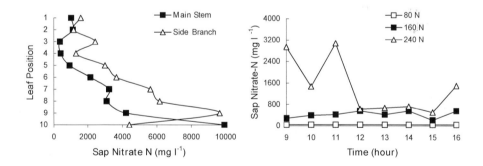

Figure 7   Nitrate-nitrogen concentration in petiole sap of 'Maris Piper' potato, as affected by leaf position (leaves counted from the top) and diurnal time. (Adapted from MacKerron et al., 1995).

Plant reaction time to addition of supplementary nitrogen

Application of nitrogen during the growing season is expected to slow or reverse the decline in plant nitrogen concentration. In most cases this actually happens temporarily (Vos & Bom, 1993; Waterer, 1997). Both the magnitude and rate of the reversion depend upon the method used and the compound analysed. Petiole nitrate-nitrogen concentration gives a larger and more rapid response to supplementary nitrogen applications than does the SPAD-value (Fig. 8). The rapid increase of petiole nitrate-nitrogen concentration after supplementary applications indicates that uptake of additional nitrogen occurred rapidly. However, despite the consistent increase in petiole nitrate-nitrogen concentration, sometimes there is no yield response to supplementary application of nitrogen; for example when the initial nitrogen supply is very scarce. In that situation there is inadequate leaf area during early growth and the canopy intercepts a lower proportion of light (see Chapter 1.2). After supplemental nitrogen application, the plant takes time to produce more leaves but it is not able to recover the energy (light) already lost. So, the best response to supplementary nitrogen occurs if the initial supply is sufficient for good leaf growth and full ground cover and if supplements stop the crop from dying early.

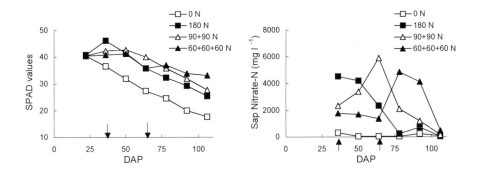

Figure 8   *SPAD values and petiole sap nitrate-nitrogen concentration in 'Vebeca' potato, as affected by different number of split nitrogen dressings. Nitrogen applied at planting (180N), and split between planting and one (90+90N) or two (60+60+60N) dressing. Arrows indicate timing of additional application. (Modified from Vos & Bom, 1993).*

Conclusions

The nitrate loss from agricultural land can be lessened by reducing nitrogen fertilization rates and by increasing the use of soil nitrogen reserves. This could be achieved by reducing nitrogen fertilization rates at planting and then applying supplemental nitrogen as needed at early- and mid-season, based on a tissue-testing program. In order to apply optimal rates of plant nitrogen, the basic information required is the relation between nitrogen supply and tuber yield (quantity and quality) and the detection of the plant nitrogen status. Results from plant analysis must be interpreted considering that concentration of nitrogen compounds decreases throughout the growing season. Environment, cultivars, crop management, and nitrogen supply all modulate this trend. Therefore, it is not possible to establish a general Critical Value or Range for nitrogen fertilizer recommendation. That information must be obtained from well conducted field trials that must take into account the environmental conditions, and the cultivars and the management strategies that are usually adopted in each specific area.

The analysis of nitrogen in the plant must be precise, quick and cheap and the stage of growth at sampling (preferably based on physiological rather than chronological age) must be accurately determined for successful interpretation. Standardizing methodology in such a way would aid and promote the use of tissue analysis for assessing and correcting the nutrient status of the crop during the course of growth. Subsequent chapters will give additional information on the management of nitrogen supply and demand in potato fields.

Recommendations

1. The estimation of the requirement of the crop for nitrogen is very important both for economical and environmental reasons.
2. Soil tests, carried out before planting, provide an estimate of the amount of nitrogen to be applied, then plant analyses can be used to monitor crop nitrogen status throughout the growing season, allowing adjustment for errors in fertilization.
3. The analysis of nitrogen in the plant must be precise, quick and cheap.
4. The best application of plant analysis depends upon the detection of a Critical Value or Range, below which the nitrogen requirement would be closely related to the deviation from it.

5. The Critical Value or Range may vary with plant part, stage of growth, cultivar, and climatic or technical factors, other than the nitrogen supply.

6. Well conducted field trials, taking into account the environmental conditions, and the cultivars and the management strategy which are usually adopted in each specific area, are needed to establish such Critical Values or Ranges correctly.

2.2   Determination of crop nitrogen status using invasive methods

N.U. Haase, J.P. Goffart, D.K.L. MacKerron & M.W. Young

Several methods exist to estimate plant nitrogen status during growth and after harvest.

During the growing season, measurements of nitrogen uptake and metabolism provide tactical information that may allow the grower to avoid loss of yield or quality of tubers and leaching of nitrogen to ground water. After harvest, determination of nitrogen in the harvested crop will allow assessment of the quality of the product and can assist planning future fertilizer application by enabling calculation of a nitrogen balance. Consumers are interested in the quality and environmental impact of potato production as well as the price of potatoes.

Invasive methods have the benefit of providing precise analytical information that allows the best decisions for future fertilizer application to be made – but which nitrogen compound is to be detected and by which method? Which methods give sufficient accuracy? Does the method needs a lot of equipment? Can you use the method by yourself? And what about the costs?

This chapter aims to answer these questions through a critical overview of existing invasive methods.

Introduction

The pressure groups involved (farmers, advisers and scientists) make different demands on the value and the accuracy of results in nitrogen assessment. Today, several analytical techniques are available, from simple procedures, which do not need experienced operators, to very complicated techniques requiring skilled technicians. Some methods offer results within a few minutes but often, analyses require several hours to complete.

Analyses with rapid and simple methods (e.g. nitrate in petiole sap, (Williams & Maier, 1990)) have been suggested to support fertilizer strategies during the growing season. However, balancing nitrogen uptake (total nitrogen in all plant parts) requires that a more extensive preparation and analysis is chosen, e.g.

Kjeldahl (Bradstreet, 1965) and Dumas method (Kirsten & Gunnar, 1983). Consumers are interested in quality aspects of potato tubers (e.g. protein or nitrate concentrations in tubers) and the environmental impact of potato production. Each of those concerns imposes its own constraints on the choice of analytical procedure.

Invasive analytical methods are well documented, tested and widespread. The depth of practical experience of these analytical procedures and their optimisation over many years makes them very reliable. In most cases interpretation of results is straightforward but this is not the case with the new and rapid invasive methods. All the non-invasive methods (subchapter 2.3) that have been developed reflect traditional analytical methods.

The use of any invasive method implies destruction of the biological structure before measurement and, therefore, further observations will require selection of other plants.

The objectives of this paper are to describe several invasive methods that are available for assessment of nitrogen concentration in the potato plant, and to outline for each one the nitrogen compounds to be detected. We will also indicate the advantages, disadvantages and field of use for each method. It is to guide users in the choice of method to support decision making in the optimal use of nitrogen fertilizers.

Description of methods

There are several methods for determining nitrogen compounds and many modifications of the basic methods have been reported. Therefore, only basic information can be given in this chapter. If more specialist questions arise that cannot be answered from the references cited at the end of this chapter then please contact the authors or national experts. A description is given to:

1   Kjeldahl-Digestion
2   Dumas-Combustion
3   Spectroscopy
4   Near-Infrared Spectroscopy
5   Chromatography
6   Ion Specific Potentiometry
7   Test Strip Procedure

Note that the methods of sampling and the preparation of samples strongly influence the quality of results, for details see subchapter 2.4.

Kjeldahl-Digestion

In 1883 Kjeldahl published his method of nitrogen determination. The Kjeldahl method is a two-step procedure (ICC (International Association for Cereal Chemistry), 1994; Young et al., 1993). First, the different oxidation- and binding-types of nitrogen are transferred to ammonium sulphate (($NH_4)_2SO_4$) by heating the organic matter with concentrated acid and a catalyst (oxygen transfer). Several acids or combinations of acids and different catalysts have been tested for special purposes. Environmental considerations have led to the introduction of new catalysts (e.g. mercury-free ones).

Second, after treatment with alkali the sample is distilled and the ammonia liberated is collected without any losses in a receiving vessel (Fig. 1). A titration procedure determines the amount of ammonia evolved. Concentrations of protein or nitrogen are calculated by multiplying this figure by a factor.

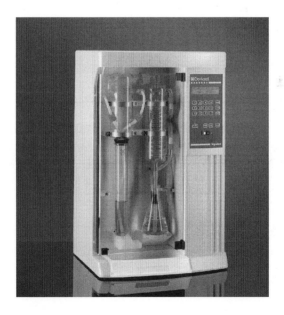

Figure 1   Distillation and titration unit of the Kjeldahl method (photo by courtesy of Gerhardt, Bonn, Germany).

The accuracy of the results depends on several aspects. Nitrogen losses occur during digestion where temperatures are above 400°C. Also, the digestion time is important. After the sample has been clarified, further boiling time is necessary. The traditional procedure does not detect all nitrogen compounds quantitatively. Where the more common catalysts are used, nitrate is only partially included. Therefore, this method gives a 'crude' protein value. However, it is possible by the right choice of a catalyst to get detailed information about several nitrogen compounds. The analytical procedure covers several single steps:

> Sampling
> Preparation (drying, grinding)
> Aliquot (2.5 g)
> Wet digestion (acid + catalyst)
> Distillation (alkaline conditions)
> Titration (neutralisation)
> Calculation

Advantages: This method is accepted world-wide and serves as a standard method. Several companies offer semi- or fully automated equipment. The information gained is on a fundamental level. Aspects of the nutritive value of tubers can be analysed.

Disadvantages: This method needs a well-equipped laboratory and experienced staff. In spite of new distillation units with automatic titration units, it is a long, drawn-out procedure. Therefore, no rapid interpretations are possible during the growing season and the method does not enable quick responses.

Field of use: Fresh or dried samples from all plant tissues may be analysed. Different N-compounds are detectable by the right choice of acid and catalyst. Estimation of the concentration of organic nitrogen in plant or soil samples where decisions are not dependent on a rapid result.

Dumas-Combustion

The dried sample is burned in an oxygen-rich atmosphere and the combustion products are purified to isolate elemental nitrogen (N_2). The quantity of nitrogen is determined by heat conductivity (Kirsten & Gunnar, 1983; AACC (Americal Association of Cereal Chemistry), 1995).

The instruments available (Fig. 2) run automatically, but the procedure is complicated and several working steps are combined. The material should be milled finely (<1 mm) to allow intimate mixing of the constituents. A weighed sample is brought into the furnace and ashed at 1100°C under an oxygen stream. The combustion gases are transported with a carrier gas (carbon dioxide <CO_2> or helium) through a second combustion area, through a sulphur dioxide (SO_2) absorption-tube, and over a copper catalyst.

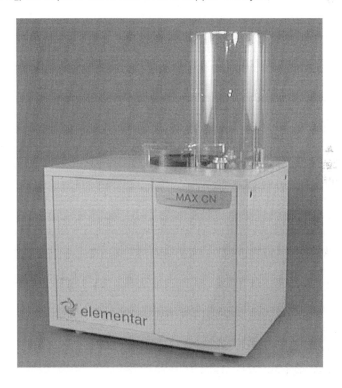

Figure 2   *Dumas-combustion unit (photo by courtesy of Elementar Analysensysteme, Hanau, Germany).*

During this sequence, oxygen is removed and nitric oxide is reduced. In the next step water vapour is condensed and the carrier gas flows through drying tubes to the heat conductivity detector:

Sampling
Preparation (drying, grinding)
Aliquot (150 mg – 1 g)
Charging the combustion unit
Automatic measurement (ashed at 1100°C)
Automatic calculation of total nitrogen content

Advantages: This method has become a standard procedure. Time for analysis is very short, and a total number of 300 samples per day is realistic.

Disadvantages: The equipment is expensive and the laboratory must have experienced staff. The relative small aliquot to be used requires an accurate pre-treatment (sampling, preparation, and homogenisation). Fresh material must be dried. As was stated for the Kjeldahl procedure, the method is not suitable if a fast reaction is required, within the growing season. The results are more applicable on a fundamental level. Differentiation between different nitrogen compounds is not possible.

Field of use: Estimation of the concentration of organic nitrogen in plant or soil samples where decisions are not dependent on a rapid result. Dried samples from any plant tissues may be analysed. Dumas combustion gives total nitrogen content.

Spectroscopy

Most of these analytical procedures are divided into two steps. First, the nitrogen component has to be extracted or to be transferred in a measurable form (by oxidation or reduction). Organic nitrogen must be transferred to ammonia; inorganic nitrogen has to be extracted from the matrix. Instead of a titration technique, specific dye binding reactions can be used for determination. This requires a spectrometer and a calibration with pure substances.

Measurements are carried out in a defined range of concentrations over which there is a linear relation between the instrument's response (light extinction) and the concentrations of ammonium or nitrate. If the concentration is too low or too high then a concentration or a dilution step must be added to the procedure. Provided that the samples are of an adequate quality, no distinction needs to be made between foliage, stems, roots, and tubers. But that is the crucial point: quality of sample. Samples are small and so, if an analysis is to represent the whole plant, sub-sampling must be done extremely carefully to avoid bias between one tissue or another. It would be better to sample and analyse each tissue separately. At least the tissues will share a common calibration.

Examples for determination of organic- and inorganic-nitrogen are described below.

Several automatic systems are available for enhanced determination (e.g. the Auto Analyser System). The first step of nitrogen determination (oxidation of reduced nitrogen and extraction of nitrate respectively) is carried out as described above. The second step of the procedure uses continuous flow systems with staining reactions. Samples are brought into the tubing system of the flow stream by an auto-sampler. Normally, air bubbles separate samples. Chemicals are added at different points. After a defined reaction time, photometric detection takes place in a continuous flow cell and the results are often processed automatically. These expensive systems run well with a high number of samples. If only a few samples are to be analysed, a more conventional procedure is to be recommended.

Organic N:

After digestion with sulphuric acid and a catalyst (see Kjeldahl digestion; oxidation) the resulting ammonia is detected through a staining reaction with hypochloride and phenol in presence of nitroprusside (Mitchell, 1972):

```
Sampling
Preparation (drying, grinding)
Aliquot (20 mg)
Wet digestion (acid + catalyst)
Dilution with water, aliquot
Chemical reaction
    -   copper binding
    -   neutralization
    -   staining
Measurement after 2h at 623 nm against
calibration with ammonia N
Calculation
```

Inorganic N:

Nitrate (normally ammonia can be neglected) has to be extracted from the samples, whether fresh or dehydrated, then an enzymatic reduction to nitrite is carried out and consumption of nicotinamide adenine dinucleotide phosphate (reduced form; NADPH) is detected twice (40 and 60 min after start of reaction) at 334, 340 or 365 nm (Roche Diagnostics, formerly Boehringer, 1997).

```
Sampling
Preparation (drying, grinding)
Aliquot (5 g)
Extraction (hot water)
Protein precipitation (Carrez I and II)
Filtration
Enzymatic reduction of nitrate
Measurement 40 and 60 min after start
of reaction at 334, 340 or 365 nm against blank
Calculation
```

Advantages: Precise and reliable measurement of different nitrogen compounds (organic/inorganic) is possible.

Disadvantages: Experienced staff is needed. The equipment is expensive, particularly in the case of automatic systems. Fresh materials or liquid extracts must be dried before analysis (in case of organic N).

Field of use: All plant parts can be analysed. Suitable where a large number of samples are to be processed.

Near-Infrared Spectroscopy

Near-infrared spectroscopy uses either the reflection of radiation in wavelengths between 1100 and 2 500 nm (9 091 to 4 000 cm^{-1}; NIR = near infrared reflectance) or its transmission in wavelengths between 800 and 1100 nm (12 500 to 9 091 cm^{-1}; NIT = near infrared transmission) (Fig. 3). An irradiated sample reflects a specific light intensity (NIR) or absorbs a light signal (NIT) (Osborne & Fearn, 1986).

The wavebands are broad and not very precise because of overlapping and overtones, e.g. combined oscillations of C-H, O-H and N-H groups.

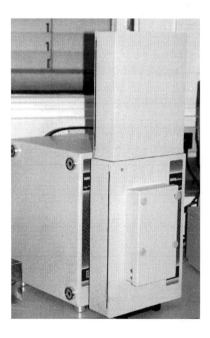

Figure 3   NIR-measuring unit (photo by courtesy of Foss, Hamburg, Germany).

NIR and NIT are secondary techniques that rely on calibration against a reference analysis (e.g. Kjeldahl or Dumas). Separate calibrations are required for each tissue type and the quality of the calibration is critical to the success of the methods.

A calibration set of samples of each tissue type (40 in a minimum) is chosen to cover the range of values expected in further measurements. The spectral data from the calibration set of samples are analysed by principal component analysis to summarize most of the variation in the data set into linear combinations of the original variables (reflectance or transmission in discrete wavebands). Multiple regression is then used to relate the scores of the principal components to the analytical values for nitrogen concentration obtained by more conventional analysis. The calibration ought to be validated using smaller sets of further, independent tissue samples on which both conventional analysis and NIR (or NIT) measurements are made. This is to be repeated periodically. After calibration of the instrument and validation, the measurements are very cheap, fast, and easy to do:

Sampling
Preparation (drying and grinding or mashing)
Representative aliquot (10 – 50 g)
Charge the NIR/NIT instrument (vessels, petri dishes)
Measurement (< 1 min)
Automatic calculation
(multiple compound detection)

Advantages: In contrast to the difficulties of choosing a representative sample set for calibration, single measurements are easy to do. Highly trained personnel are not needed. Simultaneous determination of several constituents in a sample is possible. The accuracy of the near-infrared spectroscopy technique often is very close to that of chemical techniques that it replaces. The precision of NIR/NIT is sometimes better than for conventional analyses. Established calibrations are available for use with potato (Young et al., 1997) that would require only validation measurements to be made before they could be applied.

Disadvantages: Near-infrared spectroscopy is an empirical method. There is no mathematical law that describes the interaction of radiation with a scattering medium containing a heterogeneous distribution of absorbing molecular species. Therefore, the instrument readings are arbitrary and require a calibration using samples of known composition.

The high water content of fresh potato material (leaves, stems, tubers) results in a high signal that may obscure that of other constituents. Therefore, NIR models were developed with dried and finely ground plant material (Young et al., 1997). In the case of NIR- and NIT-models with fresh (homogenized) potato material (NIR: (Büning-Pfaue et al., 1998); NIT: (Weber et al., 1996)) settlement or sedimentation of homogenized samples should be avoided. Of course, each plant tissue needs its own calibration.

Field of use: Both techniques, reflection and transmission, have been tested by potato researchers. Results are available for dry matter, starch and protein content and nitrogen concentration in different plant parts.

Future developments: Further research is needed, to decide if common calibrations are suitable for potato samples coming from different varieties, sites, and years. Developments in other areas like cereals, milk and meat industry show the advantages of calibration networks. Maybe these will help to reduce the problems of calibration.

Chromatography

Several chromatographic systems are available but here we consider high-performance liquid chromatography (HPLC), which is commonly used for analysis of many substances. In principle, dissolved compounds are separated by interactions between a stationary phase and a mobile, carrier phase containing the material to be analysed.

After a pre-treatment (extraction and clarification) samples are applied on a column that contains a specific matrix. The several compounds in the samples are moved through the column in the mobile phase, at different speeds because of differential interactions with the column matrix. Consequently, the several component substances are eluted one after another. Normally, spectroscopic determination of the relevant compounds is made in a continuous flow cell. Pure substances are used for calibration.

The HPLC-method can be used for several nitrogen components but its practical application for protein or amino acid composition is limited and, so, we consider here the benefits and disadvantages of its use for determination of nitrate only. There is an official HPLC procedure for nitrate determination (CEN (European Committee for Standardisation), 1997).

Nitrate is extracted from the matrix by hot water. Clarification is carried out by precipitation reactions (Carrez I and II) or by Solid Phase Extraction (SPE). The pre-treated samples are applied to the HPLC-instrument and nitrate is detected at 205 nm (mobile phase, flow rate and column specification are described in the cited standard procedure):

Sampling

Preparation (drying and grinding or mashing)

Representative aliquot (10 g with 15 mg nitrate in a minimum)

Extraction

Clarification

Protein precipitation (Carrez I and II) or Solid Phase Extraction (SPE)

Ultrafiltration

Injection into HPLC-system

Calibration with nitrate

Calculation

Advantages: Accuracy of results is high. Method is used as a standard.

Disadvantages: A well-equipped laboratory and highly trained staff are needed. The equipment is expensive.

Fields of use: All plant parts can be detected both fresh and dried material. Official analyses of nitrate are normally done using this technique.

Ion Specific Potentiometry

This method can be used to detect nitrate levels in fluids (Usher & Telling, 1975). An ion specific electrode (ISE) and a reference electrode detect an electrical potential between the electrode and the solution by charge separation. This charge depends on the concentration of ions that can pass through a membrane (in this case nitrate ions). The ion-dependent voltage should be measured without any current. To prevent interactions an ion equilibrium solution is added:

Sampling
Preparation
(drying and grinding or homogenizing)
Representative aliquot (2 g dry matter and 10 g fresh matter resp.)
Definite volume (for best results)
Sample aliquot + ion equilibration solution
Measurement
(nitrate selective and reference electrode)
Calibration with nitrate
Calculation

For practical measurements a calibration curve must be developed but it is generally linear and so very few concentration values may be needed (two are not enough). For practical purposes the procedure can be mechanised in a flow injection system.

Advantages: This method allows easy detection of nitrate in both fresh and dry samples. Because of the high precision of the measurement, it is not necessary to make replicate observations and, since the response time of the instrument is a matter of seconds, operation is fast. The instrument has a wide dynamic measuring range so that concentrations varying by factor 100 to 100,000 can be made without loss of the accuracy. The volume of measuring solution can be very small. Most instruments have automatic temperature compensation and they can be automated, allowing a high sample frequency. The costs of both the equipment and its operation are low.

Disadvantages: Selectivity depends on the ion to be measured. Interactions with other anions are possible leading to reduce accuracy at higher total ion loading. The size of a necessary nitrogen supplement can not be indicated.

Field of use: Nitrate determination of petiole sap and tuber extracts. The method is suitable for field measurements.

Test Strip Procedure

Test-strip procedures allow rapid estimation of plant compounds such as nitrate (MacKerron et al., 1995; Nitsch & Varis, 1991). On a small plastic strip, a pad is impregnated with an indicator that reacts selectively with nitrate to give a coloured product. The intensity of the colour corresponds directly to the nitrate concentration in the sample. For rough measurements, the intensity of the colour can be estimated visually by comparison with a graduated scale that is often provided with the test strips. More accurate information is obtained by using a hand-held photometer, specially designed for the test strips (Fig. 4).

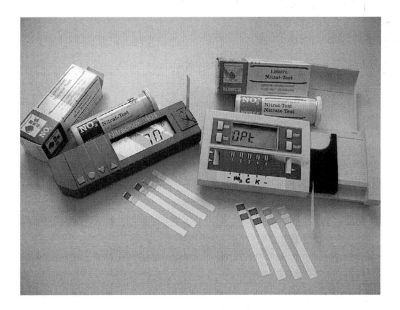

Figure 4   Test strip equipments (Test strips: Merck, Darmstadt, Germany; Nitrachek 404: Wolf, Wuppertal, Germany; RQflex: Merck, Darmstadt, Germany).

Test strips can be placed on the cut surface of potato tubers or can be used with sap of mashed potatoes or with petiole sap. Analyses are possible directly in the field but better values are given by combining sap from a number of petioles or tubers collected under a proper sampling routine (see chapter 2.4) and, if necessary, by careful dilution before testing:

> Sampling
> Preparation (cutting, homogenizing resp.)
> Representative aliquot (some liquid is enough)
> Test strip reaction
> Measurement of stained strips
> Calculation if necessary

The test strips have a low dynamic range over which the colour intensity changes linearly and so plant sap may have to be diluted carefully after a preliminary measurement. Although the calibration against chemically pure solutions of nitrate is good it has been found to be less good where plant sap is used and, worse, the calibration has been shown to differ between years (MacKerron et al., 1995). That difference may be related to the differences between packages of test strips, reported by van Loon and Houwing (1989). They also reported large variation between test strips in a single package. The first two of those deficiencies can be controlled by calibration of each package against known test solutions, repeating the test between years if a package lasts across two seasons. (One manufacturer of test strips uses a special hand-held photometer to overcome that problem: A bar code strip with a correction factor is attached to each package). Using several test strips (at least five) for each determination can control the second deficiency.

Measured values of nitrate concentration in petiole sap vary diurnally and within the plant so it is important that a proper sampling protocol is observed (see chapter 2.4).

The values obtained from the tests are compared against one or more 'reference' curves of concentrations of nitrate changing against time to suggest whether or not the crop is adequately supplied with N. The validity of such reference curves is still a matter of debate. Among the complicating factors are (i) the crop can take up luxury levels of nitrate when it is available in the soil and store it in the cell vacuoles for later reduction; (ii) low levels observed during a dry spell may be followed by a peak after rainfall.

Advantages: Values for concentration of nitrate in crop tissues are obtained on the same day as the samples are made. The photometer and the test strips are cheap, and the method is amenable to operation by the farmer himself.

Disadvantages: The accuracy and interpretation of results is often very poor. The method does not indicate the size of a nitrogen supplement that will be required.

Field of use: Estimation of concentration of nitrate in petiole sap and in tubers. Farmers and advisers can use it in the field themselves.

Conclusions

Most of the analytical methods are suitable for measuring one or more nitrogen compound. To find out the best method a two-step procedure is recommended:
1. Decide the nitrogen compound of interest (e.g. nitrate, protein)
2. Decide the accuracy of results required (simple analyses may be less informative, but may be desirable where costs need to be minimized.)

The tabulated summary (Table 1) will help identify the best method for a particular purpose. Each method is annotated with an indication of accuracy. Relative cost of analysis is also indicated, as there is no common agreement about prices. Special preparation techniques are necessary for some of the analytical methods and type of compounds of interest. Both the costs of the equipment and the labour requirements have a strong influence on the cost of analysis.

Invasive measuring techniques are widespread all over the world. Most of these methods are used as standards. Some are officially accepted by several international organisations. Within the last few years, modern automatic equipment has been developed for common procedures that allow high capacities and automatic calculation of results.

Table 1   Overview of the analytical methods for assessment of nitrogen (N) compounds in potato plants.

Method	Nitrogen compound to be detected				Quality of results	Experienced staff	Costs
	Total-N	Protein	Non protein-organic N	Inorganic N (nitrate)			
Kjeldahl Distillation	X	X	X	X	high	yes	mid
Dumas-Combustion	X				high	yes	mid
Spectroscopy		X		X	high	yes	high
Near Infrared Technique	X	X			low-high	yes/no	low
Chromatography				X	high	yes	high
Ion Selective Potentiometry				X	high	no	low
Test Procedure: Control strips				X	low-mid	no	low

If advisers and farmers want to make the measurements themselves, the only method available is the estimation of nitrate with test strips or with a nitrate specific electrode. Each of these procedures has the potential for the wrong interpretation of results by inexperienced users. In laboratories, experienced people do investigations and interpretations. Unfortunately results are not available just after sampling, so that their interest for use in quick and early assessment of plant nitrogen status in fertilizer strategy during growing season is really limited.

2.3   Determination of crop nitrogen status using non-invasive methods

R. Booij, J.L. Valenzuela & C. Aguilera

Modern potato growers want to produce potatoes in an environmentally sound way. But how can potatoes be grown with minimal losses to the environment? How do we know whether or not the crop is supplied with sufficient nitrogen during crop growth? Does the crop indicate its nitrogen status, can it be monitored and what methods are available to determine it easily?
In the following chapter a number of non-invasive methods will be described to monitor crop nitrogen status.

Introduction

Measurement of the nitrogen status of the crop is getting more important, if part of the nitrogen should be applied during crop growth in accordance with the actual crop nitrogen demand. Non-invasive techniques make it possible to assess the crop nitrogen status instantaneously in the field, so that a time lag between measurement and nitrogen supplementation can be minimized. There is no restriction to the number of measurements to carry out and measurements can take place on the same spot every time if needed. These techniques are only an advantage if the time needed for measurement is short, and they are cheaper and just as reliable as an invasive technique. The equipment needed for measurement should be easy to handle. Lack of nitrogen can be expressed in reduced leaf expansion and in altered leaf colour. Therefore, both aspects are good candidates for monitoring crop nitrogen status non-invasively during crop growth (Lemaire, 1997). Present non-invasive techniques are related to these aspects. Candidate methods should not only result in a highly distinguishable signal that is unambiguously related to the nitrogen status, but that signal should also be translated into a recommendation for the amount of nitrogen to be applied.

As the need for such systems is relatively new, none of the methods described is yet completely developed for practical use in potatoes.

Chlorophyll meters

Chlorophyll-meters indicate the chlorophyll concentration in the leaf by measuring the light attenuation at the wavelengths 430 and 750 nm. At 430 nm no light is transmitted, due to absorption by chlorophyll, while at 750 nm, the light is transmitted. Because there is a strong relation between chlorophyll concentration in the leaf and the nitrogen concentration (Fig. 1), the level of chlorophyll in the leaf is a good indicator of its nitrogen status.

The SPAD meter (SPAD 502, Minolta, Osaka, Japan) is a commercially available form of chlorophyll meter (Plate 1). The instrument is easily hand-held and up to 30 individual readings can be stored, selectively deleted, replaced and averaged. The measuring head consists of two hinged parts that clamp onto the leaf, each measurement takes only a few seconds.

Fully expanded leaves at the top of the canopy are the best indicators of the nitrogen status of the crop. Within a field the mean of about 30 leaves should be taken from plants randomly distributed in the field. (See Chapter 2.4 on sampling).

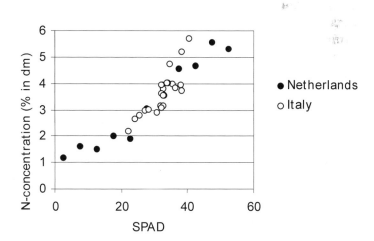

Figure 1   *Relation between SPAD readings and concentration of nitrogen (%) in the dry matter of an individual leaf. (data from Della Vedove and Vos).*

Plate 1   SPAD-502 in use (photo by AB-DLO).

There is experimental experience in using this instrument on potatoes but we do not know of its commercial use. From these experiments a preliminary recommendation has been formulated, that the SPAD-value should be higher than 40 - 41 during a period of 7 to 8 weeks from emergence (Fig. 2). The earlier- the SPAD values drops below this threshold, the higher the amount of supplemental nitrogen should be.

Chroma meter

The colour of leaves is related to the nitrogen concentration of the leaves, among other things. A high nitrogen concentration gives the leaf a deep green appearance, while a low nitrogen concentration results in a paler green or even a yellowish appearance.

With a Chroma meter, the colour of an object can be assessed objectively, by converting all colours that the human eye can perceive into numerical codes.

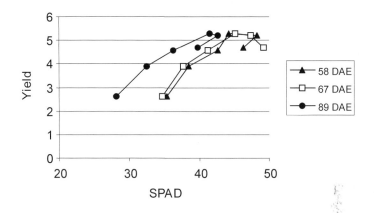

Figure 2   *SPAD readings at several developmental stages (days after emergence (DAE))*
               versus final tuber yield (data from Colauzzi).

The most widely accepted standardised (CIE) parameters to define colour are 1
(tightness), a (redness- greenness), and b (blueness – yellowness). Positive
values correspond to red and yellow while negative values correspond to green
and blue.

Colour can be measured by a portable Minolta Chroma Meter CR-200 (Minolta,
Osaka, Japan), a compact tristimulus colour analyser for measuring the reflective
colours of a surface. The Chroma meter has an 8 mm diameter measuring area
and uses a viewing angle of 0°. A pulsed xenon arc lamp in a mixing chamber
provides diffuse, even illumination over the sample surface. Six highly sensitive
silicon photocells, filtered to match the CIE (Commission Internationale de
l'Eclairage) Standard Observer Response, are used by the meter's double-beam
feedback system to measure both incident and reflected light.

As with the chlorophyll meter, experience in using colour as an indicator for the
nitrogen nutrition of potatoes is restricted to experiments. Information is still
lacking about optimal colour development during potato growth, how to
accommodate varietal differences, and how to respond by applying additional
nitrogen.

This also means that there are still no recommendations as yet. An indication of
the effect of nitrogen availability on the colour parameters is tabulated here
(Table 1).

Table 1   Chroma meter readings and crop nitrogen status.

Nitrogen status	Parameter	
	a (red-green)	b (blue-yellow)
Low	< -40	> 30
Normal	< -40	10-20
High	< -50	-10 -10

To establish the nitrogen status of the crop, the colour of the terminal leaflet (20 plants) from the youngest mature leaf should be determined. (See Chapter 2.4 on sampling).

A cheaper method, but based on the same principle is the use of the Munsell colour cards. Chromatic colours are divided into five principle classes, namely red, yellow, green, blue and purple. Combinations of the five principle classes, give intermediate classes such as yellow-red, etc. The following classes can be distinguished for nitrogen in potatoes (Table 2).

Note of caution: The ability to judge colour differences, using the colour cards, varies from person to person and consistency should be learned in practice. In diagnosing nutrient deficiencies, it should not be forgotten that the colour of the plant tissue may be influenced by other factors besides its nitrogen nutrition. These factors include shortage or excess of water, low temperature or incidence of diseases. Moreover, the colour manifestation of nutrient deficiencies is not always the same across species. Therefore several years of experience are needed to use the method for nutritional diagnosis.

Table 2   Color classes and crop nitrogen status.

Nitrogen status	Hue name
Low	2.5 GY (814 to 8/1o)
Normal	2.5 GY (4/4) or 5G (514 to 5/8)
High	2.5 BG (3/6) or 5 BG (312 to 316)

Chlorophyll fluorescence

Chlorophyll fluorescence in a leaf depends on the ability of the photo-system to transfer light into chemical energy. Environmental stresses such as excessive irradiance, drought, chilling, herbicides and nutritional stresses affect this ability. As nitrogen deficiency affects chlorophyll and the proteins of the photo-system, chlorophyll fluorescence can be used as an indicator for nitrogen stress of the leaf. But is also modified by other stresses.

There are two characteristic parameters of chlorophyll fluorescence to consider, namely F_o (basic fluorescence level) and F_m (maximum fluorescence after high illumination). The ratio of $(F_m - F_o) / F_m$ is proportional to the quantum yield of photosystem , reaching a value of 0.75-0.85 in unstressed plants. Nitrogen deficiency causes an increase in F_o and a decrease in F_m.

The equipment needed is portable (Plant Efficiency Analyser, HANSATECH, King's Lynn, UK) and a measurement can be made within one minute. About half an hour before the measurement a clip should be put on the leaves to be measured, because leaves need to be acclimated in darkness for at least this period of time.

Again, only limited experience is available in potatoes and only from experiments and there are no recommendations available yet on critical values. However, some tendencies can be given. When leaf is stressed the ratio $(F_m - F_o) / F_m$ is less than 0.80. However, that is only a general indicator of stress. To make it more specific for the measurement of the nitrogen status, one has to take advantage of the nitrogen gradient within a canopy (high nitrogen concentration at the top and low at the bottom). Chlorophyll fluorescence should be measured on five successive, mature leaves on a stem and the gradient in $(F_m - F_o) / F_m$ gives an indication of the nitrogen status. The crop suffers from lack of nitrogen where the slope of the relationship between the ratio (after logarithmic transformation) and leaf position on the stem exceeds 0.012.

Light interception, crop reflection

Light interception

Both methods are basically related to the one process, namely the development of leaf area. For potato, as for most crops, there is a linear relation between total nitrogen uptake and leaf area index (Lemaire, 1997; Fig. 3), until maximum leaf area is reached.

Light interception by the canopy is closely related to the development of the leaf area index. Commercial hand-held equipment is available. The Sunscan (Delta-T Devices, Cambridge, UK) uses a 1-metre long probe set with an array of sixty four probes to sample the distribution of photosynthetically active radiation along a transect of the crop floor. This has an immediate advantage over spot sensors, in that the instrument samples the spatial heterogeneity of the plant canopy. The instrument is supplied with in-built software that allows repeated measurements to be stored and then averaged. It is good practice to make a set of ten recordings for each average value that is stored. The number of average values to be recorded will depend on the uniformity of the canopy but should not be less than ten, i.e. a total of one hundred measurements. This number can be reduced if experience shows that the average values do not vary much. The in-built software converts field measurements into estimates of leaf area index based on direct and diffuse incident light, transmitted light, solar zenith angle, leaf angle distribution, and the absorption of photosynthetically active radiation by leaves (Campbell, 1985; Norman & Jarvis, 1975). Where some of the variables are not available the leaf area index can still be calculated from the ratio of sunflecks to diffuse light under the canopy.

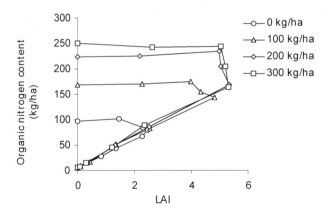

Figure 3   Development of leaf area (LAI) and nitrogen taken up during crop growth at four N
            application rates.

Another example is the LiCor 2000 (LiCor, Lincoln, USA), that also measures light interception quickly and that converts measured values into LAI. Intercepted light is calculated from measurements above and below the canopy. Within a field twenty measurements are needed to obtain a good average. A calibration line is needed to convert LAI-values into total crop nitrogen content (Fig. 3).

Crop reflection

The third and most sophisticated method is a measurement based on light reflection by the canopy (Plate 2).
In comparison with bare soil, the reflection in the near infrared increases with development of the canopy, while reflectance in the red decreases. So reflection in these two wavebands can be used to estimate leaf area and, as shown before (Fig. 3), leaf area is strongly related with total crop nitrogen uptake. Additionally, leaf colour is strongly affected by nitrogen concentration,

Plate 2   Use of cropscan in potatoes (photo by AB).

as indicated by the chlorophyll concentration in the leaf. In fact, canopy reflection provides an estimate of the total amount of chlorophyll per unit area that is, in turn, closely related to nitrogen concentration. In this way, light reflection by the canopy can be related to nitrogen concentration in the leaves.

For crop reflectance measurements a CROPSCAN (Cropscan Inc, Rochester, USA) can be used. With this instrument the hemispherically incident radiation and the radiation reflected by the canopy in upward direction can be recorded simultaneously and reflection is represented as percentage of the incident radiation. Filters are installed to delimit the wavelength bands that are measured. The viewing angle of the system and its height above the crop are set so that an area of about 1 m^2 is measured. The equipment is easy to carry and each measurement takes only a few seconds.

There are some important pre-conditions, however, before the instrument can be used:

- Solar elevation angle must be greater than 30°
- There must be no shadows on the area to be measured
- The canopy should be dry
- If it is windy, the number of measurements must be increased.

The method is still under development, so experience is limited to experiments.

Overview of performance

Several non-invasive methods are available to estimate crop nitrogen status, but all are more or less still in a developmental stage for use in potatoes. The best is, of course is the one that informs best about crop nitrogen status, however, a comparison has not been made yet. But also other aspects determine the potential for use of which several are given in Table 3.

Table 3   Performance of the different non invasive methods.

Method	Aspect					
	SPAD	CM	CC	CF	LI	CR
Stage of development	++	+	+	--	+	+
Time window for application	++	++	++	++		++
Sensitivity environmental conditions	-			++		+
Labour needed to apply			+	+	++	+
Skills needed to apply	-		+			-
Robustness of equipment	+	+	+	+	+	
Price of equipment			--	+		+

SPAD: idem; CM: Chroma meter; CC: Colour cards; CF: Chlorophyll fluorescence; LI Light interception; CR: Crop reflection (++ more than average; -- less than average).

Conclusions

In this chapter a number of non-invasive methods have been described to determine the nitrogen status of potato crops. All the methods are more or less related to estimates of the amount of chlorophyll present in the canopy, a plant characteristic closely related to nitrogen content. Although not all methods are fully developed yet, radiometric measurements such as SPAD and crop reflectance are the most promising. Both methods are able to estimate the crop nitrogen status and would allow timeous supplementation of nitrogen to correct any deficiencies. However, in common with the invasive methods described in Chapter 2.2, these methods do not, of themselves, give an estimate of the size of the supplement that will be required.

The methods described here share a second limitation that none is a specific indicator of the nitrogen status of the crop. That is, none is uniquely influenced by nitrogen supply. This does not disqualify the methods for use but shows that each must be interpreted with care.

Recommendations

1. To reduce nitrogen losses, part of the nitrogen should be applied during crop growth in accordance with crop nitrogen status.
2. Radiometric methods can be used to monitor crop nitrogen status.
3. Strenuous efforts are needed to develop such methods fully.

2.4   Spatial and temporal aspects of sampling of potato crops for nitrogen analysis

J.P. Goffart, M. Olivier, D.K.L. MacKerron, R. Postma & P. Johnson

When potato plants are to be assessed for N, either for consultancy or research work, some basic questions must be answered first. What should be sampled - whole plant or specific plant organ? When to sample - optimal time within the growing season and within the day? What is the optimal pattern of sampling, and optimum sample size? And how should the samples be handled, stored and sub-sampled before analysis? This chapter aims to answer all these questions for most common invasive and non-invasive methods as described in Chapters 2.2 and 2.3.

Introduction

In the framework of optimizing the use of nitrogen in potato production, the main purposes of collecting field potato plants or plant organs for analysis of their nitrogen concentration can be summarized as:
(i) estimate the nitrogen concentrations in the crop; and directly or by inference:
(ii) estimate the total nitrogen uptake by the crop.

These estimates may be used for:
(iii) assessing the requirement for supplementary application of nitrogen fertilizer during the growing season.
(iv) forecasting yield (quantity and quality).

If the purpose of the measurements is to estimate the requirement for supplementary nitrogen - (iii) above - then, although the whole plant can be used, it is more usual to sample specific plant organs. These are then tested by a rapid method in order to give a « quick » and early indication of the total plant or crop nitrogen status. Such methods usually involve measuring the concentration of inorganic nitrogen in plant tissues, for example, the petiole sap nitrate concentration (see Chapter 2.2), or an indirect measurement of the concentration of total nitrogen, and are carried out in the field on the crop. In this case non-invasive methods are often used, based for instance on an assessment of leaf chlorophyll concentration. Although these methods are being used increasingly, they still have some problems and are not yet completely developed for practical use in potatoes (see Chapter 2.3).

Where (ii) is of interest, it is necessary to sample the whole plant - leaves, stems and tubers. Analysis for total nitrogen is performed in the laboratory with standard, accurate analytical techniques such as Kjeldahl analysis or Dumas combustion or, more recently, Near Infrared Reflectance (NIR) (see Chapter 2.2). Then measurements of plant growth, or dry matter production are also necessary.

As explained in Chapter 2.1, the concentrations of nitrogenous compounds in whole plants and in plant organs vary both in space and time, at field and plant level, through a variety of causes. For instance, the concentration of total nitrogen in dry matter is normally the highest in leaves intermediate in stems and the lowest in tubers (Kolbe & Stephan-Beckmann, 1997). This is the reason why, when total nitrogen uptake is required, it is necessary to analyse the tissues separately (see later).

These sources of variation must be considered when sampling plants in the field. In addition, special procedures are required for both the pattern of sampling and sample size to give adequate spatial coverage: optimal sampling at the field level is different from sampling at the scale of an experimental plot.

The objectives of this chapter are to give a short description of the sources of variation in nitrogen concentration of a potato plant, and then to explain the optimal procedures for sampling potato plants in space and time and how to handle and store the samples. The procedures differ slightly for assessment of the concentration of total nitrogen in whole plants by Kjeldahl or Dumas methods, the petiole sap nitrate test and the hand-held chlorophyll meter test. It is important to observe these differences because spatial and temporal sampling critically influence the values of nitrogen concentration that are determined and consequently contribute greatly to the quality and usefulness of data in decision making.

Sampling techniques (whole plant or plant parts)

All analytical methods give a measure or an estimate of the concentration not the amount of nitrogen or its compounds so, if an estimate of the amount of nitrogen taken up is required, it is necessary also to measure the plant production, either fresh weight or dry weight, or both. That requires a certain minimum amount of plant material as will be shown later in this chapter.

Sources of variation in potato plant N

Table 1 summarizes the main sources of variation in plant nitrogen concentration at each spatial scale of sampling. At field and plot levels, the spatial variation of soil nitrogen supply within a field (see Chapter 3.5.) has a strong influence on the observed concentrations of plant nitrogen compounds. The uptake of nitrogen varies also between plants, even if they are grown in a common set of conditions. Within a plant, nitrogen concentration may vary according to plant tissues, stem position or leaf position (see Chapter 2.1.). Within a plant organ, the position of the part being considered (e.g. a leaflet on a leaf or part of petiole or part of tuber) can also induce variation in nitrogen concentration.

Temporal sources of variation are summarized in Table 2. The stage of development of the crop is a main source of variation as soil nitrogen supply and the kinetics of nitrogen uptake and plant growth fluctuate greatly during the growing season (see Chapter 2.1.). For any nitrogen compound, e.g. nitrate, the concentration within a plant organ may vary within a day due to mineral fluxes and metabolism - nitrate is reduced to ammonium and so to amino acids - within the plant tissues. Care must be taken to avoid dehydration of the samples while they are being collected, handled, and stored, as this would cause an apparent increase in plant nitrogen concentration expressed on a fresh weight basis. Finally, weather conditions can also induce important variation in nitrogen concentration.

Table 1   *Sources of spatial variation reported in the literature for concentrations of*
              nitrogenous compounds in potato at several scales.

Sampling level	Range of sampling area unit	Sources of variation
Field	Several thousands square metres	N supply (soil and fertilizer). See Chapter 3.4. N plant uptake. See Chapter 2.1
Plot	Several square metres	N supply (essentially soil) N plant uptake
Whole plant	About 1/4 square metre	Plant tissues (leaves, stems, tubers). See later in this chapter. Stem position (main or lateral) Leaf position on a stem (old or new). See later in this chapter.
Plant organ	About 1/1000 square metre	Leaflet position on a leaf 'True' petiole or including midrib of leaf Position within a tuber

As a corollary, the choice of the plant organ and the form of plant nitrogen to be checked are of prime importance for the accuracy and relevance of the assessment of nitrogen status. In addition, the higher the values of nitrogen measured the lower the risk of significant errors in measurement. Two general rules can be considered in the choice of and time for plant organ to be sampled:

1) As future growth rate is linked to the availability of N, the best plant part to sample is that showing the greatest range of variation in nitrogen concentration whilst nitrogen supply goes from sufficient for optimal growth rate to deficient.

2) If nitrogen deficiencies are to be rectified then they must be detected early so that timely decisions can be made on the application of supplementary nitrogen fertilizer.–Leaf growth makes heavy demands on the supply of nitrogen and is necessary for both full canopy expansion, and duration of a good canopy throughout the planned life of the crop.

Table 2   *Sources of temporal variation in plant nitrogenous compounds concentration.*

Stage of development of the crop	(seasonal variation in both available N in the soil and in uptake by the plant, and plant growth)
Time of the day	(diurnal variation, effect of temperature and solar radiation)
Handling and storing conditions of the samples	(light, temperature, humidity)

Inherent variation in growth and production of potato

As mentioned in the previous section, where the uptake of nitrogen is to be estimated, it is necessary to assess the amount of growth made by the plants but that growth is variable, even under closely uniform conditions. Between neighbouring plants, initial small differences can increase as the plants develop. An early emerging plant tends to be larger than a later emerging one. A plant growing in a wider space tends to be larger than one crowded by immediate neighbours. When a sample of plants is taken, some will be larger than others, some smaller. The average or mean value can be calculated and also the standard error of the mean (SEM). The SEM gives information about how good the mean of the sample is as an estimate of the true mean of the population of all the plants in the field. A useful indicator of the uncertainty in the estimated mean is provided by the coefficient of variation (CV = SEM / the mean).

Figure 1 illustrates how the CV of total dry weight and tuber yield, measured in the cultivar Maris Piper, decline with increasing sample size. Table 3 shows how that translates into meaningful guidance for a sampling strategy. For example, if the true difference between two treatments (say, some nitrogen and more nitrogen) were 10%, then one would need to take twelve samples of six plants or four replicates of nine plants in order to have a 90% chance of detecting the difference.

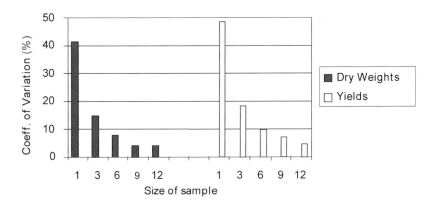

Figure 1   *Coefficients of Variation (%) of plant dry weight and tuber yields in potato as*
                functions of sample size (number of tubers per sample).

Sampling for assessment of total nitrogen by invasive methods: Kjeldahl, Dumas, or NIR

Total nitrogen concentration is usually determined on whole plant dry matter. The plants must, however, be separated into leaves, stems and tubers. As stated earlier, the concentrations of nitrogenous compounds differ between leaves, tubers, and stems. Since these tissues are also present in differing amounts, it is extremely unlikely that any aggregated sample would be a representative one. More detail on sub-sampling is given later in this chapter. The standard analytical methods for total nitrogen analysis are the Kjeldahl digestion and Dumas combustion but the analytical procedures of both these methods are relatively slow and expensive so that these methods are more likely to be used for research work than in consultancy or commerce. Near Infrared Reflectance (NIR) has also been used for the same purposed as a faster, cheaper and sufficiently accurate method (see Chapter 2.2.). However, any system for determining the requirement for supplemental applications of fertilizer, that relies on estimating total uptake of nitrogen by the plant will suffer from the labour intensive and time-consuming procedures in plant sampling and sub-sampling. That approach is, however, the best for research work. As the

Table 3   Number of replications required for a 90% chance of finding a significant difference
(P = 0.05) using each of five sample sizes (number of plants per sample).
(MacKerron, unpublished data).

True difference (% of mean)	Size of Sample				
	2	3	6	9	12
5	~	~	45	12	12
10	> 50	35	12	4	4
15	31	35	12	4	4
20	18	9	4	2	2

concentration of total nitrogen in dry matter of leaves, stems and tubers decreases with progressive development (see Chapter 1), it may be necessary to take sequential samples during the growing season, depending on what ones wishes to do. If it is intended to assess the nitrogen status or nitrogen uptake of the crop then sampling should be started at 100% crop emergence and further samples should be taken every 10 to 15 days during the growing period. Samples should be taken shortly before complete senescence or destruction of the haulm, and an extra sampling of tubers and dead stems at harvest is also recommended if possible.

When sampling plant for sequential as well as final harvest in a field or an experimental plot, take care to harvest all plant material (stems and tubers) and collect all fallen laminae from the sample area but keep them separate from the living parts.

Sampling for assessment of inorganic nitrogen by invasive methods:

Petiole sap nitrate test

In this case, sampling focuses on a particular plant organ: the petioles. For the potato crop, leaves and petioles are probably the most useful organs for rapid assessment of nitrogen status during the growing season, if one cannot sample the whole plant. The petioles collected from the newest fully expanded leaves are the best to use, usually the 4th or 5th from the top of a main stem (Jones & Painter, 1974) (Fig 2a). Immature leaves at the top of the plant and older, partially shaded ones lower on the plant should be avoided.

The main reasons for this choice of leaf and tissue are related to several sources of variation in petiole sap nitrate concentration and to general rules mentioned before. First, the concentration of nitrate in petiole sap is five to ten times greater than in leaflets (Goffart, unpublished data) and it is relatively easy to extract the sap. Second, it has been demonstrated (see Chapter 2.1) that the concentration of nitrate in petiole sap increases with depth into the canopy in petioles of both main stems and side branches (Millard & MacKerron, 1986; MacKerron et al., 1995; Biemond & Vos, 1992). Tests at the end of July showed that on the main stem the concentration of nitrate ranged from 440 ppm in the top leaf to 1760 ppm in the 9th leaf from the top (Millard & MacKerron, 1986). For most leaf positions, the concentration of nitrate is much higher in side branches than in main stems (MacKerron et al., 1995). Finally, the lowest values and the lowest variation in the concentration of nitrate in petiole sap are found at leaf positions 3 and 4 on the main stem, confirming these leaves as being the best to use for this test.

The concentration of nitrate in petiole sap changes with time and the developmental stage of the crop, normally decreasing but, possibly increasing after either the supplemental application of fertilizer nitrogen or rainfall after a dry period (MacKerron et al., 1995) (see also Chapter 2.1).

Consequently, it is best to start the sampling at around tuber initiation (usually 20 to 30 days after 50% crop emergence and to take further samples every 7 to 10 days, during the growing period, for another six to eight weeks, depending of the maturity class of the variety.

The time of day also has an influence on the nitrate level in petioles, for the concentration of nitrate is usually higher and more variable in the morning than in the afternoon, especially if there is ample nitrogen. The best time for sampling is from midday to mid-afternoon if one is uncertain of the nitrogen supply. Where one is confident that the nitrogen supply is low, it may be acceptable to sample from late in the morning until mid-afternoon (MacKerron et al., 1995) (see Chapter 2.1.). Sampling between 8 a.m. and noon has also been reported (Williams & Maier, 1990) but that is a time when the concentration of nitrate is variable and may be temporarily high, suggesting that the crop is well-supplied with nitrogen when it is not. As a rule, sampling and testing should always be at the same time of the day, but certainly not early in the morning or late in the afternoon.

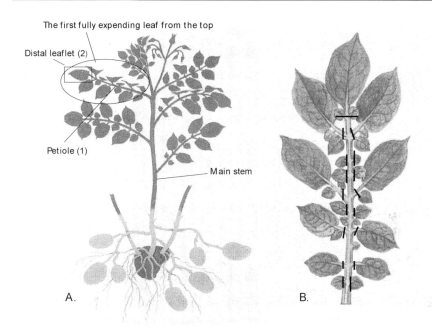

The first fully expending leaf from the top

Distal leaflet (2)

Petiole (1)

Main stem

A.   B.

Figure 2   *A. Sampling procedure on plant for the petiole sap test (1) and the hand-held*
               chlorophyll meter test (2); B. For the petiole test, remove all the leaflets (large and
               small) from the petiole before crushing it. (Illustrations of potato plant and leaf
               redrawn from Lab. Morphog. Exp. Végét., Orsay, Paris, France).

The preferred size of sample and sampling pattern is described later in this chapter. Once collected and prepared (Fig. 2b), the petioles can be crushed between a glass plate and an inert roller such as a plastic pen. The sap can then be obtained and analysed.

Sampling to assess the concentration of total nitrogen by non-invasive

methods: Hand-held chlorophyll meter test

The SPAD meter (SPAD 502, Minolta, Osaka, Japan) is one commercially available form of the hand-held chlorophyll meter; another, recent development is known as the HNT meter (Hydro N-tester, Hydro-Agri Europe, Brussels, Belgium). At present, they provide the most developed, non-invasive method for estimating the nitrogen content of foliage (see Chapter 2.3). For descriptions of sampling strategies for other non-invasive methods such as the Chroma meter, chlorophyll fluorescence, ground cover, light interception or crop reflectance, refer to Chapter 2.3.

Using the SPAD (or HNT) meter, sampling is focused on the leaflet. It is best to use fully expanded leaves at the top of the canopy (Fig 2a) as those are fully illuminated and contribute most to crop growth. Normally the distal leaflet of the 4th or 5th leaf from the apex of a main stem is used for a single measurement with either chlorophyll meter (Vos & Bom, 1993).

Sampling should start at around tuber initiation, and further measurements should be taken every 7 to 10 days during the growing period. There is no evidence available to suggest that one time of day is better or worse than any other.

Note that when using a hand-held chlorophyll meter, it is important to avoid taking measurements on abnormal, yellowish leaves. Further, care should be taken to avoid recently applied pesticides on leaves (especially fungicides), water from dew or from rainfall, and dust, all of which can interfere with the measurement. Also avoid the midrib of the leaflet.

Pattern of sampling and sample size in commercial field for advisory or consultancy work

General recommendations

The main problem is to ensure an adequate spatial coverage of the field. In the Netherlands, the maximum area recommended to be represented by plant sampling is two hectares. In the UK, ten hectares is used as the maximum.

First of all, a visual inspection of plants in the field is helpful. Where parts of the field evidently differ from others (yellowish leaves, smaller plants, infestation by weeds, etc.), it is better to sample from them separately; and to keep the samples separate.

Where more than one cultivar is grown in a field, treat them separately, as cultivars can have different nitrogen concentrations even if grown in the same conditions (see. Chapter 2.1). Finally, avoid headlands, tramlines, wheelings, overlaps of sprayers and especially the former sites of old manure heaps.

Guidelines for whole plant sampling

Sampling whole plants in a commercial field to provide a timely recommendation of requirements for fertilizers may be considered to be unrealistic in commercial practice as it is labour intensive and time consuming and the analyses, too, take time. However, there are a few recommended patterns of sampling within a field that are applicable to sampling soil, whole plants, or plant parts as they aim to cover the whole field. The method shown here may not be wholly justified from a statistical point of view (distribution of variability in the concentration of total nitrogen between individual plants within the field) but, where any apparent differences are small, it is the best that is available.

A recommended way of sampling across a field to ensure adequate spatial coverage of the field is illustrated in Fig. 3, based on the classical W pattern. It is recommended in France for potato crops (Comité National Interprofessionnel de la pomme de terre, CNIPT, Paris , France). Other, similar procedures are recommended elsewhere.

The procedure is as follows: before starting, estimate the length and breadth of the field to determine the number (n) and length of individual sampling areas to be taken and the required interval between each individual sampling area (cf. formulae in Fig. 3). If the field is homogeneous, consider the recommended number (n) of individual sampling areas as a function of the field area, as stated in Figure 3. The recommendation is to sample whole plants along at least one metre from one row in each individual sampling area starting more or less in the middle of the area.

Figure 3 illustrates the example of a field with an area of three to five hectares, where the recommended number of samples (n) is twelve. Collect whole plants from twelve planting rows, divided in four series of three rows. The first three rows are sampled while travelling in one direction, the following three rows in the opposite direction, and repeat the same for the remaining six rows. Start harvesting whole plants from the first individual sampling area, and go to the following area by skipping several rows for the distance indicated in Figure 3.

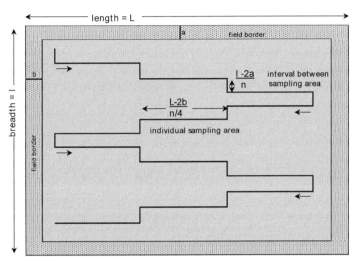

n – number of individual sampling areas (figure illustrates the situation for n=12)

Recommended values for number of samples (n) as a function of field area	
field area (ha)	n
0 to 1	4
1 to 3	8
3 to 5	12
5 to 10	16
more than 10	4 per 5 ha

Figure 3   Recommended pattern of sampling for whole potato plant or plant organs based on the classical W pattern and sampling pattern recommended in France for whole potato plant.

Guidelines for sampling plant organs (petioles and leaflets)

We consider that twenty petioles per hectare is a minimum number for adequate representation of variability, and that sample could usefully be increased to forty per hectare. The field should be traversed in the way shown in Figure 3, and no more than one petiole should be taken from any one plant.

Considering a field of two hectares, as an example, the petioles and attached leaflets would be collected from four series of two rows (n = 8). To collect a total of forty petioles, pick five petioles from each of the eight individual sampling areas (each petiole from one individual plant), then cross to the next area by skipping several rows as indicated. Immediately after collecting the leaves, remove the leaflets from the petiole (Fig. 2b). For subsequent samples, try to avoid using the same plants.

A similar pattern of sampling is recommended for assessment of the nitrogen status using the hand-held chlorophyll meter (SPAD or HNT). The HNT technique requires at least thirty individual readings per hectare (one reading per plant). For a field area of two hectares, for instance, at least eight individual readings should be taken in each of the eight sampling rows or individual sampling areas as described for petiole sampling. The same plants may be used for subsequent measurements.

Guidelines for measurement with the crop reflectance method

(see Chapter 2.3)

Measurement of crop reflection should be done on eight different places equally distributed within a field of 2 ha. On each spot take six different measurements in a half-circle around the position taken in the field. Take always stand facing the sun to prevent your shadow falling on the area being measured.

Pattern of sampling and sample size in experimental field plots (research)

Guidelines for measurements on whole plants

Four plants are probably sufficient for assessment of nitrogen concentration, but insufficient for an estimate of nitrogen uptake because of the higher variation in biomass production between single whole plants. As shown in a previous section, the minimum size of sample is related to the number of replicates taken. The coefficient of variation approaches a low asymptotic value where four replicates of eight or nine plants are sampled. That level of replication represents the most efficient gauged by the total number of plants taken.

The design of the experimental plots should provide for at least two discard or guard plants at each end of the plot and also between plants taken in sequential harvests. The plots should also incorporate two guard rows on either side.

It is sometimes customary to take larger samples at a final harvest but there can be little justification of this in reduction in the coefficient of variation.

In all cases, it is recommended that all plant material grown within the sampled area is harvested (stems and tubers); collect all fallen laminae from the sample area and separate it from the whole plant.

Guidelines for plant organs (petioles and leaflets)

There is little statistical evidence in the literature for any particular standard sampling procedure for assessment of petiole nitrate concentration. Reported sample sizes range from eight through fifteen to twenty petioles per plot, with differing levels of replication. Sample one petiole per plant. Although we recommend that subsequent samples should not be taken from the same plants, we recognise that, depending upon the frequency of sampling, the plot size may present a limitation. If a plant is to be sampled more than once, then it should be an absolute rule not to sample the same stem as before.

For measurements with the SPAD version of the hand-held chlorophyll meter, twenty readings per plot (one per plant) have been common practice but with the HNT version thirty readings are used to give a mean value.

Comments on the variability of the measurements

There are two underlying problems to be considered. One is that the plants to be measured are liable to be variable, even when grown in uniform conditions. The sizes of the samples and the sampling pattern that have been recommended here are designed to accommodate that variation, or include it within the measurement. The samples may then be said to be representative of the field or plots.

Care with sampling procedures is important so that the coefficients of variation (CV) of the concentration of total nitrogen in dry matter between individual plants, of nitrate concentration between petioles, or of chlorophyll concentration between leaflets will be as low as possible for the accuracy of the measurement. The second problem to be faced is that the site to be monitored, whether field or an experimental plot, will include variation in resources – the supply of nitrogen and water. Where that variation is obvious, one can stratify the sampling – sample each area separately. Then one can, ideally, take separate decisions for each type of area. However, a more common condition is that the variations are continuous and not obvious. Then this second source of variation becomes important as merges with the first source of variation and the consequent 'average' values represent neither one condition nor another and it is more difficult to make an effective decision on any corrective action that might be necessary. The prospect of using the techniques of 'precision farming' may offer a means to resolve that particular problem (See Chapter 6.2).

It is interesting to compare CV (= standard error of the mean divided by the mean) for values of nitrate concentration in petiole sap and of hand-held chlorophyll meter measurements at the scale of experimental plots.

For measurements using a hand-held chlorophyll meter, the CV for individual readings within a plot ranges typically between 5 and 9% during the first two months after emergence (for n = 20); late in the season the CV increases to 10 to 13% (6). Values of nitrate concentration from petiole sap tests appear to have higher CVs than HNT measurements (Table 4, Goffart and Olivier, unpublished data). The latter show very low values for CV and also for Least Significant Difference (LSD) (p=0.05) between the means of nitrogen treatments, leading to significant differences between nitrogen treatments. The CV of nitrate concentration in petiole sap is seen to increase considerably over the growing season while, for HNT values, the CV increases slowly and remains below 5% until the end of July. Such variation in the values from the petiole sap test poses a problem for the application of the technique at a commercial field level.

The use of non-invasive methods such as crop reflectance (see Chapter 2.3), measuring larger samples of plants (areas of at least one square metre) will probably generate lower CV values between similar samples and may allow the grower to address the problem of the second source of variation – spatial differences across the field.

The effect of variation across a field has been discussed already. It might be possible to take an intermediate step between traditional agriculture and site-specific agriculture (see Chapter 6.2) by following the recommended pattern of sampling for whole plant or plant organ as already described and by keeping

Table 4   *Averages of mean values and of Coefficients of Variation (CV%) of Petiole Sap*
            nitrate (PSNC) and Leaf Chlorophyll Content (LCC) generated in experimental
            plots of potato (Goffart and Olivier, unpublished data).

Methods	Statistics	Time (days after planting and date)				
		70	82	91	103	117
		12/6	24/6	3/7	15/7	29/7
PSNC (ppm)	average of mean values	6563	5114	4054	1226	621
	average of CV %	6.9	9.8	13.5	65.9	133.9
LCC (HNT values)	average of mean values	612	579	541	508	479
	average of CV %	2.7	2.1	3.9	4.6	4.9

Measurements were made on test strips with a Nitrachek photometer for petiole sap nitrate
concentration (PSNC) and with an HNT hand-held chlorophyll meter. The data are presented as
a function of time (days after planting), for nitrogen treatments differing in the total amount of
nitrogen fertilizer and whether or not the nitrogen application was split. Single mean values are
from four replications (blocks). Each replication includes the mean values of twenty petioles for
PSNC and of thirty readings for HNT. Data collected in Gembloux, Belgium, in 1997, with
cultivar Bintje.

the samples from each individual sampling area separate. This procedure offers
the possibility of roughly mapping the variation in the nitrogen concentration of
plants within the field, and then to take action in optimizing nitrogen application
according to the variation. If laboratory analyses were required, this would be
prohibitively expensive but, where a cheap and quick assessment could be
made on site, a grower might consider the approach. Then the sampling
procedure could be modified to reflect the eventual passage of the fertilizer
boom.

Handling, storing and sub-sampling (for invasive methods)

Guidelines for handling whole plants

Haulm and tubers can be placed together in clearly labelled polythene bags but should be kept separate from any fallen laminae that have been collected. Try to keep tubers and the basal ends of stems together at one side of the bag so as to minimise the amount of earth on the leaves.

At final harvest, collect up all plant material (mainly dead stems) from the harvest area and place it in clearly labelled polythene bags. Lift the tubers using a harvesting machine or fork and collect by hand. Place top or stolon material exposed by this process with the stem sample already collected.

After intermediate and final harvests, if the plant material is not to be processed immediately, it should be stored at 4°C until it can be washed, sub-sampled and prepared for the laboratory.

At intermediate harvests, when stems and laminae are to be analysed separately, it will greatly ease the labour of sub-sampling if the individual stems are ranked in order of size and the middle eight stems are chosen. The fresh weights of the whole sample and the sub-sample should be recorded. The laminae can then be stripped from the stems.

At any harvest, tubers should be sub-sampled with regard to their size distribution. Determine, on a fresh weight basis, the proportion of tubers in each size grade. Then, to provide a sample of fifty tubers, for example, multiply each of the proportions by fifty to give the number of tubers to be sampled from each grade. Where tubers are to be cut, they should be cut longitudinally to avoid bias.

Guidelines for plant organs (petioles and leaflets)

For the petiole sap nitrate test, place the petioles in a black plastic bag (well labelled and closed) immediately after collection. Keep the bag in a cool box until it is taken to the laboratory, and then store it in a refrigerator at 4°C until the sample can be processed.

No particular procedure is recommended for sub-sampling petioles.

Information to be collected while sampling

It is of prime importance to collect the following information before leaving the field:

- General performance and vigour of the crop (i.e. growth stage, presence of heterogeneous canopy areas in the field, ...)
- Detectable incidence of insects or disease damage (in all or parts of the field)
- Soil conditions (moisture, dryness)
- Climatic conditions during sampling (occurrence of rain during sampling period)
- Date of sampling and time of day when sampling was done
- Potato cultivar, crop growth stage (and date of emergence if available)

Conclusions

Recommended procedures have been presented for sampling potato plants for assessment of nitrogen concentration or total nitrogen uptake in either the whole plant or in plant parts, for the petiole sap nitrate test and the hand-held chlorophyll meter test (SPAD or HNT). Some of the details differ between full-scale field sampling and in experimental plots. These are summarized in Table 5. Proper sampling is critical for the value of the analytical data, and, therefore, must be considered as an integral part of the laboratory work. Otherwise, subsequent analytical work is a waste of time and money.

Table 5   *Tabulation between methods of analysis for nitrogen concentration in the potato plant and recommended sampling procedures for the most common methods used in practice.*

Sampling characteristics	Methods for analysis of nitrogen concentration		
	Kjeldahl/Dumas	Petiole sap nitrate test	Hand-held Chlorophyll meter test
N compound to be detected	total nitrogen in whole plant	nitrate	total nitrogen in leaf
Plant part to be sampled	whole plant must be split into leaves/stems/tubers	petiole of the 4th/5th leaf counting from the top of a main stem	distal leaflet of the 4th/5th leaf counting from the top of a main stem
Time of sampling	every 10 to 15 days starting at 100% emergence, ending at haulm senescence or destruction. Extra sampling at harvest is useful	every 7 to 10 days starting at 100% emergence, ending 6 to 8 weeks later	every 7 to 10 days starting at 100% emergence, ending 6 to 8 weeks later
Pattern of sampling in field (consultancy)	W pattern	W pattern	W pattern
Sample size in field (consultancy)	n times 1m long in a row (n as a function of field area)	at least 20 petioles per ha (one petiole/plant)	at least 30 readings per ha (one reading/plant)
Pattern of sampling and sample size in experimental plots	For sequential harvest, 8 to 9 plants in each of 4 replicated plots in two or four adjacent central rows. For final harvest, Same procedure except if larger samples are required for tuber analyses	20 petioles per plot (one petiole/plant)	30 readings per plot (one reading/plant)

Recommendations

1. For assessment of plant nitrogen concentration or nitrogen uptake in potato crops, do always follow standard recommended procedures for sampling whole plants or plant parts. Collect the right plant part at the right time (within the day or the growing season) for the specific nitrogen compound to be assessed.

2. Do always follow specific sampling procedures for both pattern of sampling and sample size to give adequate spatial coverage. Remember that optimal sampling at the field level is different from sampling at the scale of an experimental plot.

3. For invasive methods, do not leave the freshly collected plant material exposed to light or at room temperature. Store it immediately in cool temperature conditions until laboratory analysis.

2.5 Perspectives for use in practice - How can assessment of plant and crop nitrogen status be used in practice?

D.K.L. MacKerron

The farmer should aim to grow potato crops in which the supply and demand for N are in balance, so as to limit losses of N to the environment, and yet to maintain yield levels. There are several related issues that face the potato grower: What will be the crop's requirement for N and the supply of N from the soil, how effective will any application of manure or fertilizer be and, therefore, how much should be applied? The influence of water supply on the uptake of N on can have serious consequences for the environment. Inadequate supply of water can leave mineralized N, principally nitrates, liable to leaching in autumn and winter and over irrigation causes leaching to the ground water. It is important that any system that is intended to address the issue of optimizing the use of either resource should consider both.

The purpose of this part of the manual is to show how estimates of the N-status of the crop can be used to modify fertilizer practice in real time, during the current growing season.

Principles

The first estimate of the total requirement of a crop for nitrogen must be based on an estimate of the yield that is anticipated from the crop at harvest (Chapter 2) Essentially, larger crops require more nitrogen and smaller ones require less. The farmer then must make an assessment of the amount of N that is likely to be supplied by the soil. There are traditional methods available, often rules of thumb, based on soil-type and previous cropping history, and there are more analytical methods available. These several methods are assessed in Chapter 3 of this manual. Earlier sections of this chapter (Sub-chapters 2.2 - 2.4) have indicated how one can check the nitrogen nutrition of the crop as it grows.

There is so much uncertainty in the whole operation of growing a crop that a farmer is unlikely to make the correct decision at planting time on the nitrogen requirement of any one crop. With care and the best advice, he or she might even be correct on average but that means that sometimes the application will have been too high and sometimes too low.

Also, it should be evident that where too great an application has been given, there is nothing that can be done. The crop will grow with an excessive supply of N. Its maturity will be delayed: Chapter 2. 1 . There will be difficulties in killing the haulm. The tuber quality may be lower than it should be (Chapter 1.2). And there is the near-certainty of pollution of ground water by leaching after the crop has been harvested (Chapter 1.2)

In contrast, where an insufficient amount of nitrogen has been given to the crop, supplementary applications can be given to complement the first application plus the nitrogen taken from the soil and, where supplementation is timely, there need be no yield penalty.

The requirement is therefore, for a system in which the farmer will apply nitrogen fertilizer, following organic manure or without it, at some level that is reasonably likely to be less than the crop's eventual requirement but that will be adequate for early growth. The grower should then monitor the nitrogen status of his crop using suitable plant measurements, in the expectation that it will require supplementary applications.

For good commercial practice, the farmer should choose a method of crop assessment that is reliable in its measurement, and in the interpretation that can be put on the measurement, and that provides an answer quickly. How soon is soon enough? A method that takes one to two weeks to provide an answer is completely impractical. At the other limit, a method that gives an answer within minutes might be fine but it is unnecessarily fast. The assessment should allow the farmer to make a decision and implement it within a period of only a few days. One day would be ideal, three days quite acceptable.

The usefulness of an analytical test is determined, also, by whether the result of the measurement tells the farmer to give no more nitrogen if the values are within a certain range or higher. Alternatively, if the values are below a critical threshold, he knows to take action and he knows that the lower the value, the stronger and more urgent his action should be. Some of the critical values have been given in Chapters 2.1 and 2.4. These threshold values change during growth, especially where the crop has been given all the nitrogen at or close to planting time. Then the grower has to know what sort of variations in nitrogen concentration would be expected in the range of values found in a well cultivated the crop at a given stage in the growing season. Again, values have been given in Chapters 2.1 and 2.4.

Features of the analytical methods

The methods available to assess crop N-status range from those that a grower might use for himself to those that require laboratory facilities and that could be used by a grower working through an analytical service. Although it is unlikely that the grower will chose which analytical techniques are to be used by a laboratory, the consideration of timeliness may constrain that laboratory. The comparison of techniques given in Chapter 2.2 should help advisors in approaching one or another analytical laboratory, or in choosing which technique to request where a choice is available.

Neither Kjeldahl nor Dumas is suited to rapid results. Chromatography is suited to critical analyses but not for routine farm use. NIR, where available, offers answers cheaply and quickly, with good precision. It also has the possibility of analyzing components of the sample; for example, concentrations of nitrogen, tuber dry matter, starch, and proteins.

The methods suitable for field samples include ion specific potentiometry, for the analysis of nitrate, nitrate test strips and chlorophyll meters. Test strips measure nitrate concentrations in sap cheaply and very quickly. Their principal drawback, however, is the difficulty of interpreting what the values mean. Chlorophyll meters have the advantages of rapid use and quick answer coupled with stability in the value. - It doesn't change in the course of a day. This method and petiole sap testing requires careful and extensive sampling to be at all meaningful (Chapter 2.4).

Sampling

Where laboratory analyses are to be used, a grower or agronomist should sample the field extensively in an attempt to get a good spatial average but he must recognize that the few results from the laboratory, inevitably, give him limited information. They tell what is the 'average' condition of the crop. Even if the average shows a satisfactory level, there may be areas of the crop that are deficient in nitrogen and others that are in surplus. For this reason a farmer should consider using methods that can show the spatial variation across a field. If nothing else, that may direct him to sample in a more detailed way within such areas of the field. It would also allow him to manage parts of the field in individual ways. Division of a field or crop into 'management blocks' would represent a considerable advance in many situations. In this way the farmer would approach the ideal position that is often not achievable without resorting fully to precision agriculture (Chapter 6.2).

If a farmer uses the simpler analytical techniques by for himself, the cost of the analysis is largely the time devoted to making the measurements. These lend themselves to repeated sampling so enabling the farmer to follow the development of the crop and to detect any deficiencies that might arise. Better still, they make realistic the option of managing a crop in such a way that it will probably develop a deficit, as the farmer will know that he can detect a deficiency and then correct it. Chapter 2.4 has described how sampling should be done for each of several purposes. It covered number of samples, size of samples, and frequencies of sampling. The chlorophyll meters (SPAD or HNT) are shown to have less variability than other methods.

Remote sensing, even from a few metres away, where the viewing area is one square metre or more, may couple rapidity of measurement with large-scale sampling. - The biggest single source of variation is still likely to be across a single field (see later chapters) and so any method is advantageous that allows an assessment of spatial variation and repeated sampling; that is, any method that the farmer can use himself whether sap test, chlorophyll meter, or crop scanner.

Figure 1 (Chapter 2.1) shows that the nearer the initial application has been to the 'correct' amount for the crop, the later any differences appear between it and the correct treatment. Therefore, a grower must expect to sample his crop repeatedly, as early sampling may not detect sub-optimal fertilization.

When samples are taken repeatedly, the farmer should beware of scaling time by days after planting as the time from planting to emergence varies with location across Europe and with weather (growing season), and seed condition. It is better is to consider timings related to 50% emergence or, better still, to a development stage such as tuber initiation.

Features of crop responses

At the very least, a good measurement should indicate whether the crop is adequately supplied with nitrogen at the present. Ideally, it should also indicate whether a crop is likely to become deficient in nitrogen in the near future; but this is not easy as it requires knowledge of how the nitrogen level within the plant will fall if the supply from the soil slows or ceases. That requires some sort of model of crop growth and the associated requirement for nitrogen, which is not currently available to farmers. On the other hand a method that is cheap and is easy to use, e.g. use of a chlorophyll meter, can minimize those problems by allowing frequent monitoring, followed by early corrective action (see Table 1). The earlier sections of this chapter have given a reasoned assessment of the value and reliability of the methods that are currently available to assess nitrogen status. They have also indicated the practicalities inherent in the methods. Some of these have been summarized here in Table 1.

Table 1.   A guide to the practicalities of assessing crop nitrogen status in the growing crop.

How soon differences in concentration can be seen	Chapter 2.1, Figure 1
Effect of soil and applied N on chlorophyll	Chapter 2.1, Figure 3
Effect of previous crops on sustained N supply	Chapter 2.1, Figure 4
Effect of water supply on indicated chlorophyll	Chapter 2.1, Figure 5
Relation between early chlorophyll and relative yield	Chapter 2.1, Figure 6
Ease of interpretation, chlorophyll v. petiole sap	Chapter 2.1, Figure 8
Overview of analytical techniques	Chapter 2.2, Table 1
Overview of non-invasive estimation	Chapter 2.3, Table 3
Sources of spatial variation	Chapter 2.4, Table 1
Sources of temporal variation	Chapter 2.4, Table 2
Relative variability in samples	Chapter 2.4, Figure 1
Number of samples to take	Chapter 2.4, Table 3
Pattern of sampling for a representative value for the field	Chapter 2.4, Figure 3
Comparison: measurements of chlorophyll and petiole sap	Chapter 2.4, Table 4

Several practical conditions, such as the use of organic manures and the occurrence of water-stress, may increase the need for good monitoring, limit the scope for action, or reduce the need for supplementary nitrogen.

The importance of the previous crop is in its effect on the sustained mineralization of organic N. For example, a preceding legume crop can lead to enhanced mineralization and so a lesser requirement for supplement while the residues from a preceding cereal crop can cause immobilization of N and result in a greater need for supplementation (Chapter 2.1). Where organic manures are used, the transition from the rapidly available mineral N in these manures to the nitrogen that is slowly released by the mineralization of the organic matter is difficult to predict (See Chapter 3). In both these cases, only frequent crop monitoring as already explained can show the correct course of action.

The apparent response of a crop, especially one grown on light soils, to the highest doses of nitrogen reflects the higher losses from those soils when all the N is given at planting time. Properly managed split applications of nitrogen will give equally high yields with lesser applications of N (Chapter 2.1).

If a crop cannot be irrigated then, in many growing seasons, the growth of the crop will become limited by lack of water, not by lack of nitrogen. Although the uptake of nitrogen is likely to be reduced, the growth of the crop, and so its requirement for nitrogen will be reduced further. Splitting the nitrogen application allows the total application of nitrogen to be adjusted. Monitoring the crop in these circumstances will almost certainly reveal high concentrations of mineral nitrogen in the plants. It is almost certainly not worth making the measurements until a water supply has been restored as the results can be foreseen, and the farmer's energies will be better spent otherwise than considering giving supplementary nitrogen (Chapter 2.1).

It is sometimes claimed by the suppliers of fertilizers that additional nitrogen can compensate for a shortage of water but this is not so. Where a crop would be deficient in nitrogen then extra N could enhance both root and shoot growth, possibly increasing the amount of water that can be extracted from the soil. But, where a crop has sufficient nitrogen during early growth this will not apply.

Conclusions

Estimates of the total requirement for nitrogen must be based on anticipated yield. The thresholds shown in Table 2 (Chapter 2.1) are for measurement values if all of the fertilizer is given at planting but they may be taken, also, as working figures where most of the fertilizer is applied at planting. Figure 6

(Chapter 2.1) indicates that the prospect of low yields at harvest is not detectable in chlorophyll meter values until after tuber initiation or shortly after that (60 days after planting in the data shown). The period for crop monitoring and effective corrective action, leaving the crop time to respond, is roughly four weeks after tuber initiation. The exception to this is with foliar application where small supplements can be given often.

None of the methods to analyse crop nitrogen status indicates the size of the supplement that should be given when a deficiency is found to be present or is imminent. There are three broad responses to the problem. The first is to give a supplement little and often, as is done where a foliar-applied supplement is given, and to continue monitoring the crop. Another approach, where less than the expected full requirement was given at planting is, simply, to apply the part withheld. That is the approach used in the AZOBIL method (see Chapter 5). The third approach is to relate the supplement to the length of the growing season remaining and the required increase in yield. That requires sophisticate decision support (Chapter 5.3).

Recommendations

The recommendations can be summarised in the following diagram:

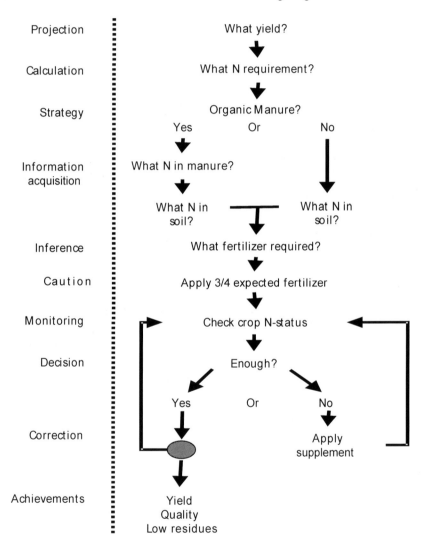

3   Soil nitrogen status

3.1   Forms of soil nitrogen

M.A. Shepherd & R. Postma

Both the amount and form of nitrogen in soils affects potato nutrition. Organic and mineral nitrogen forms are present in soils, and there is a continual interaction (or cycling) between the different fractions. Crops take up nitrogen in mineral form (ammonium and nitrate nitrogen), and it is important to understand what affects the size of this mineral N pool through the season. So, to aid better fertiliser decisions we need to know: typical amounts and forms in soils? Inputs and outputs of nitrogen to and from soils? Factors affecting conversion between forms? Variations between soils? Variations during and between seasons?

Introduction

Nitrogen is continually cycling through the environment (see Figure 5 in Chapter 1.2). Soil holds much of this nitrogen (estimated at 1.5×10^{11} tonnes in the world's soils) and therefore has an important role in cycling (Jenkinson, 1990).

Within the soil itself, there is continual internal cycling, with conversion from organic to inorganic forms and *vice versa*. Many of these changes are biological processes mediated by the soil flora and fauna (Jarvis et al., 1996). Therefore, factors that affect soil biological activity also affect rates of nitrogen transformations in the soil.

The soil receives nitrogen from the atmosphere (in rain or by dry deposition, from fertiliser and/or manure and/or plant residue additions). Nitrogen can also be fixed from the atmosphere by legumes. Returns of plant residues and manures, and application of fertiliser also contribute to soil nitrogen. The soil loses nitrogen by removal in harvested crop, by leaching of nitrates or by volatilisation of ammonia gas. Denitrification (conversion to nitrogen gas or its oxides) can also occur.

Because crops require inorganic N ('mineral N' or 'Nmin') for growth, the influence of N cycling on the soil mineral nitrogen supply is important to crop production. Variation in soil mineral N through the season can be large: this is why it is important to measure the amounts of soil nitrogen (see Chapter 3.5). Here, we aim to describe forms of soil N.

The soil nitrogen cycle

Figure 1 summarises the main forms of soil N, the transformations that take place between the different forms and the agents of these changes.

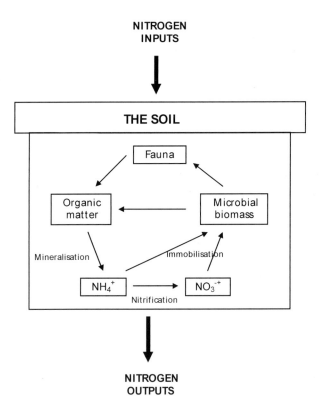

Figure 1   Representation of the soil nitrogen cycle.

The soil biomass

The release of nutrients from organic matter is dependent on the decomposition processes brought about by a wide range of soil organisms (Beare et al., 1995): bacteria, fungi, microfauna (e.g. protozoa, nematodes), mesofauna (e.g. mites, Collembola) and macro fauna (e.g. millipedes, termites and earthworms), collectively known as the 'Soil Biomass'. The biomass is at the centre of the internal N cycle (Jenkinson, 1990), as we show in Figure 1. It serves several important roles (Jarvis et al., 1996):

- an agent of change, decomposition and release of N from fresh and old organic matter
- a major 'sink' for mineral N
- a potential source of mineral N.

Therefore, most soil N passes through the biomass. The biomass itself also represents a substantial amount of soil N, typically 50-100 kg ha^{-1}.

Another term that is used is 'Soil Microbial Biomass'. This is defined as the living part of the soil organic matter excluding plant roots and fauna larger than amoeba. However, we should clearly not consider transformations solely as a microbial process because a whole range of invertebrates are involved. Importantly, numerous studies have shown that soil management systems affect soil biological diversity and activity and that much of this activity influences soil nutrient cycling processes (Lampkin, 1992), so that there will be differences between soils.

As a result of the cycling within the soil, there will be large fluctuations in mineral N during the season. Figure 2 shows a typical mineral N profile for a potato crop grown in North West Europe, with the changes that occur during the growing season.

Soil organic nitrogen

Organic matter conveys many physical properties to soil (e.g. water holding capacity, workability). Organic matter is also an important soil nitrogen pool. Most N in soils is in organic form so that the rate of its conversion to mineral N is an important contributor to the crop's nitrogen supply.

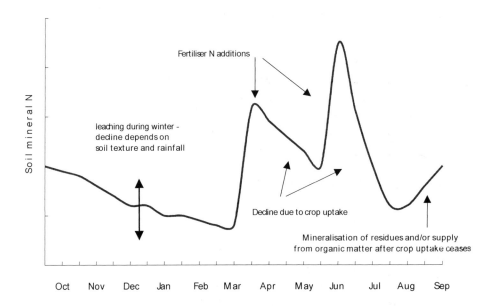

Figure 2   Representation of soil mineral N levels before, during and after the potato crop.

Composition

Organic N is composed of a continuum or organic material stabilised against further degradation to different degrees by physical separation from the soil microbial biomass and/or direct association with inorganic ions and clay surfaces (Hassink, 1992). This continuum can be divided into pools in which materials behaving similarly are grouped together. Therefore, we can sub-divide soil organic matter into many categories. For example, laboratory techniques include separation by chromatography, electrophoresis or by density, with the aim of determining which fractions are most important in terms of decomposibility.

However, for fertiliser advice, a simple but effective solution is to consider only two components:

- old soil organic matter (partly decomposed previously and then stabilised in the soil)

- additions of fresh organic materials (e.g. plant matter, organic manures). Thus, every soil has an inherent (background) N supply from old (native) organic matter, and additional pulses of N from fresh material.

Sources of fresh organic matter include the return of crop residues after harvest or by senescence during the growing season, incorporation of green manures or cover crops before planting the next crop, the application of waste materials (ranging from farm manures and wastes, sewage sludges to by-products from industrial processes) and roots and root exudates. Fresh organic materials can therefore either continually be returned in small amounts throughout the growing season (e.g. by leaf drop or root exudation), or in larger pulses (e.g. after incorporation of a cover crop or harvest residue, or after application of manure).

Amounts of soil organic N

The background levels of native soil organic matter depend on soil type (texture and drainage status), long-term cropping or other history (i.e. return of residues), topography and climate. Under most circumstances, organic matter accumulates or declines only slowly. Uncultivated soil with natural vegetation tends to have most organic matter and long-term grassland generally accumulates more than long-term arable soils. Disturbing the soil by cultivation

Table 1   The average topsoil organic N contents of arable sites in the UK (Shepherd et al., 1996).

Topsoil texture	No. of sites	Organic N (t ha^{-1} to 25 cm)		
		mean	range	
clay	7	11.7	8.0	-13.6
clay loam	34	7.5	5.0	-12.0
sandy clay loam	6	6.4	4.3	- 9.0
silty clay loam	17	5.8	4.0	- 8.6
silt loam	5	8.6	6.3	-12.6
sandy loam	31	4.7	2.7	- 7.6
loamy sand	6	4.3	3.7	- 4.9

causes oxidation of organic matter and levels decline more rapidly. The rate of decline depends on soil type, cropping and management, particularly frequency and type of cultivation, all of which will influence the balance between inputs of residues and depletion by mineralisation.

Agricultural soils contain large but variable amounts of organic matter and nitrogen (e.g. Table 1). On average, the soils contained about 7 t ha^{-1} organic N compared with only 76 kg ha^{-1} mineral N measured in the autumn after harvest.

Organic nitrogen transformations: mineralisation/immobilisation

Organic and inorganic nitrogen pools continually interact (Fig. 1). The process of (gross) immobilisation and (gross) mineralisation occur simultaneously but, in most circumstances, the net effect is the release of mineral nitrogen ('net mineralisation') into the soil for crop uptake or other processes such as leaching. Only when carbon-rich, low nitrogen, residues are added to the soil do we generally see net immobilisation or 'disappearance' of mineral nitrogen.

The rate of net mineralisation is a key factor in determining soil N supply to a crop (or available for loss processes). It is mediated by the soil biomass, as described earlier and, again, the most simple model of turnover is a slow background supply from old, stable organic matter and a quicker 'pulse' from fresh organic additions.

The amounts of mineral N released vary with soil type, residue input and, because it is a biological process, environmental factors. Nitrogen released therefore varies between soils, between fields and between seasons. Typical rates may be 0.1 - 1.0 kg ha-1 per day. However, there is no simple relationship between, for example, total N and nitrogen mineralisation though it may act as a guideline to likely release. Clearly, an estimate of soil nitrogen supply through mineralisation is necessary for fertiliser recommendations - see 3.5.

Mineral nitrogen

'Mineral N' is the term used to describe NO_3^- plus NH_4^+. Many factors affect the soil's mineral N content (Fig. 1). It is the most important N pool for crop uptake, and so we must aim to optimise its supply to the crop by matching soil mineral N release, fertiliser additions and crop uptake. Large amounts of mineral N present in the soil at times of low or no crop demand are at risk of loss from the soil - bringing a financial and environmental penalty.

Amounts of mineral N

In all soils mineral N is a small proportion of the soil's total N. Most mineral N is in NO_3^- form because NH_4^+ is quickly nitrified in arable soils. Consequently, most nitrogen is taken up by crops as NO_3^-.

Nitrate is highly mobile, with all in the soil solution, and this makes it especially prone to losses by leaching when water moves through the soil profile. Correct timing of fertiliser applications is therefore most important. Ammonium is adsorbed to negatively charged clay particles and so is less mobile. Retaining nitrogen as NH_4^+ by blocking the nitrification process with inhibitors is one method of reducing leaching losses (Chapter 3.3).

The main sources of mineral N in soils are (a) N released by mineralisation of organic matter and (b) fertiliser inputs. For potatoes, the amount of mineral N in the soil in the spring has a major influence on fertiliser requirements (see Chapter 5). Soil mineral N in spring will be derived from:
* the amounts from the previous harvest (mineralisation of crop residue and soil organic matter plus wasted fertiliser)
* any released by mineralisation during winter and early spring
* any losses, especially by leaching.

Thus, mineral nitrogen levels are likely to decrease most over winter on sandy soils and least on heavy textured soils such as silts. Measurement of soil mineral N in spring provides a 'snapshot' of the plant available nitrogen and this can be used as a basis for fertiliser recommendations (see Chapter 5). This approach is especially useful because even after the same crop, mineral N amounts can vary greatly between fields depending on previous crop management (Table 2).

Table 2   Typical levels of mineral N in soils in the autumn as affected by previous crop
             (average of fields sampled 1993-1997 in the UK).

Previous crop	No. of fields	Mineral N to 90 cm (kg ha^{-1})		
		mean	min	max
wheat	220	95	23	429
barley	146	87	14	290
oilseed rape	63	116	42	237
peas	12	115	73	181
beans	21	86	45	188
potatoes	47	111	23	354
sugar beet	48	54	11	256

Mineral N pools: Nitrification/denitrification

Nitrification is the oxidation of NH_4^+ into NO_3^-. It is a microbiological process which is carried out by a limited number of groups of bacteria and which takes place under aerobic conditions. The process can be subdivided into two steps, i.e. a conversion of NH_4^+ into NO_2^-, and the conversion of NO_2^- into NO_3^-. These two steps are carried out by different bacteria and, in general, the conversion of nitrite into nitrate proceeds much faster than the conversion of ammonium to nitrite, thus preventing the accumulation of nitrite in soil.

Important factors controlling nitrification are:
• the supply of ammonium. This will be derived either from mineralisation of organic matter as described earlier or from fertiliser additions.
• pH. Most observations indicate a lower limit for nitrification of pH 4. Combinations of high ammonium concentrations and high pH (>8) may inhibit nitrification due to the toxic affects of free ammonia (NH_3).
• oxygen supply. Oxygen will be limiting to soil nitrifiers at higher soil moisture contents, higher temperatures and/or high contents of oxidisable organic matter.
• temperature. Nitrification slows to almost zero below 4°C. The optimum temperature for nitrification appears to vary widely among soils, but the suggested range is 20-40°C.

Denitrification is the reduction of NO_3 or NO_2 via several intermediate products into the gases nitrous oxide (N_2O) and dinitrogen (N_2). The microbiological process takes place under anaerobic conditions (absence of oxygen), and it can be carried out by many different groups of bacteria. We should try to avoid soil conditions that encourage denitrification because the process represents a loss of nitrogen from the soil. Sometimes, losses will be an order of magnitude less than, say, losses by leaching. Nevertheless, since one of the products of denitrification is nitrous oxide, a greenhouse gas, this must be avoided. In some circumstances, such as heavy rainfall after fertiliser application in spring, denitrification might be as important as leaching for N losses (Addiscott & Powlson, 1992).

Important factors controlling denitrification are:
- organic carbon. Denitrification in soil is strongly dependent on the availability of organic compounds as electron donors and as sources of cellular material.
- oxygen. Denitrification in soil increases with decreasing oxygen content. The absence or reduced availability of oxygen is required for both synthesis and activity of the enzymes involved in denitrification.
- nitrate. At relatively high nitrate concentrations denitrification in soils is independent of nitrate concentration. However, concentrations below approximately 40 mg l^{-1}, denitrification kinetics appear to be first order.
- pH. Generally denitrification activity is small at pH <4.
- temperature. Denitrifiers are adapted to and capable of growth over a relatively wide soil temperature range.

Mineral N pools: ammonia volatilisation

Loss of NH_4^+ as NH_3 is another process whereby soil mineral nitrogen is lost from plant uptake. Ammonia losses from soils are generally small because (a) NH_4^+ is adsorbed to the soil and (b) it is quickly nitrified so that there is little available for volatilisation. High pH is also required. Losses will occur from manures applied to the soil surface and, sometimes, urea fertiliser. These processes are discussed elsewhere.

Conclusions

Mineral N (NH_4^+ plus NO_3^-) is the important nitrogen form for crop uptake, with much of this present as NO_3^-. However, most soil N in is in organic forms (>95%). Soil texture and previous management history affect the amount of organic N in soils, but it is usually several tonnes per hectare. There is continual internal N cycling between mineral and organic forms within the soil. The soil biomass, encompassing all soil invertebrates, facilitates this cycling. The net release of mineral N from the organic fraction (net 'mineralisation') represents nitrogen available from the soil for crop uptake or loss to the wider environment (by leaching, denitrification, volatilisation). Although there are many different fractions of organic N, it can best be considered as three 'pools': old organic matter, recent additions of fresh material and the soil biomass. Rates of soil N supply depend on many factors but are typically in the range 0.1-1 kg ha^{-1} per day: there is no simple method for predicting the release from a particular soil. For advisory purposes it is often of benefit to measure soil mineral N status as the basis to providing a fertiliser recommendation.

Recommendations

1. Soil nitrogen supply derives from mineralisation of old organic matter and additions of fresh material (e.g. crop residues, organic manures). Therefore fertiliser recommendation systems need to estimate this input.
2. Soil mineral N status at the start of the growing season is affected by soil type, climate and recent management. Therefore measurement of soil mineral N is a good starting point for fertiliser recommendations.
3. Mineral N, present in the soil at times of no or low crop demand, is liable to losses to the wider environment. Therefore management practices must aim to match soil N supply, fertiliser applications and crop N uptake.

3.2   Role and value of organic matter

G. Hofman & J. Salomez

'Soil organic matter has, since the dawn of history, been the key to soil fertility and productivity' (Allison, 1973).

Though during ages organic matter[1] was considered the key factor to soil fertility and productivity, recent agricultural (fertiliser) practices fail to appreciate the role of (soil) organic matter. So to re-value the interest of organic matter, we need to know if and how it can be adequately measured, what is its importance for soil fertility and consequently potato production, both quantitatively and qualitatively, and how it evolves under different farming conditions. Does an optimal level of soil organic matter at all exists and if so, how can this level be obtained and maintained? What are the different sources of fresh organic material? When and how must they be applied and what are the repercussions on mineralisation or the building up of soil organic matter? Practical evidences on the influence and importance of organic matter on potato crop are given in this chapter.

Introduction

As generally agreed, stable manure exerts a positive effect onto potato production, whereas slurry negatively influences tuber quality. These contradictory fundamental ideas about basically the same product point out the complexity of using organic matter when cultivating potatoes. Besides the above mentioned products, other organic (waste) products, each with their specific properties, are also in use.

The objectives to apply organic fertiliser are manifold:
* keeping up physical soil conditions:
  ◊   aggregate stability;

---

[1]   Unless otherwise mentioned, organic matter both constitutes soil organic matter and fresh organic materials

◊   soil air and soil water regime;
◊   workability of the soil;
- keeping up the soil organic matter content through the addition of fresh organic materials (farmyard manures, crop residues, green manures,...);
- a substantial direct supply of plant nutrients (N, P, K,...), and
- a prolonged supply of nutrients through mineralisation of the more stable fractions of organic matter.

To cope with all these objectives requires a good insight into all processes related to organic matter. The objective of this chapter is to give an overview of the determination, characteristics, functions, types, effects and levels of organic matter in soil, focused on potato crop.

Determination and characterisation of organic matter

The most common method for the direct determination of soil organic matter (SOM) is ignition at high temperature (950°C) (Nelson & Sommers, 1982). However, besides the destruction of SOM, loss of structural water by clay minerals and the destruction of $CaCO_3$ into CaO and CO_2 are responsible for weight losses. This means that in soils containing large amounts of hydrated alumino-silicates and/or $CaCO_3$ the organic matter content will be overestimated. To overcome these constraints a lot of modifications have been proposed, mostly including pre-treatments with different acids (e.g. HCl, HF,..) and ignition at lower temperatures (400-500°C).

Due to difficulties encountered by direct determination of SOM, indirect procedures, mostly wet oxidation methods, were proposed, expressing soil organic carbon (SOC) content as a measure of SOM. The most common used wet oxidation method, proposed by Walkley and Black (1934), is based on the reduction of $Cr_2O_7^{2-}$ in an acid environment by organic compounds and subsequent determination of unreduced bichromate by back-titration of Fe^{2+}. This method only recovers ±75% of the total SOC content, because temperature (90°C) at which the reaction occurs is not high enough to totally oxidise SOC, meaning that the real SOC content = $C_{W\&B}$ *4/3. As with the direct methods, also here modifications were proposed, mostly affecting temperature (Anne, 1945; Springer & Klee, 1954).
To convert the %SOC to %SOM, a so called 'carbon-factor' was introduced. But as the proportion of C in SOM is highly variable, different factors have been

proposed. Nowadays mostly 1.724 (Van Bemmelen factor) and 2 are in use as 'carbon-factor'. But to avoid discussions, it's more appropriate to mention SOC instead of SOM.

One further difficult step is the characterisation of SOM. It is mostly based on the partition into nonhumic (carbohydrates, proteins, peptides,...) and humic substances (humic acids, fulvic acids and humines), but this is just a purely chemical approach. More practically oriented fractionation methods link composition of SOM with mineralisation (Van Soest, 1963; Stevenson, 1965), but up till now characterisation of SOM remains merely empirical. The same fractionation methods have, on the contrary to SOM, been successfully used for the characterisation of plant residues, linking in situ N mineralisation with specific plant fractions (De Neve et al., 1994).

Functions of soil organic matter

Influence on soil physical characteristics

Soil structure and aggregate stability

Crust formation and erodibility of the soil is the consequence of an insufficient stability of soil aggregates. The aggregate instability and crust formation is correlated with the organic matter content of the soil. This is illustrated into Table 1, giving the effect of different inputs of organic material in a long term experiment (10 years) on SOM content, aggregate instability and crust formation. The potato crop is particulary sensitive to soil structural problems.
The influence of SOM content on soil erosion is illustrated in Figure 1, giving the erodibility (K-factor) for two silt loam soils with respectively 1 and 3% SOM. The K-factor diminishes from 0.47 to 0.36.

Soil air and soil water regime

As potatoes are very sensible to water shortage (see chapter 4) the upgrading of water availability will guarantee more stable and higher yields. Organic matter, including both SOM and plant residues, may exert a positive effect on

Table 1   *Influence of %SOM on aggregate instability and crust formation in silt loam soils*
               (Hofman, 1977).

	Object[1]			
	1	2	3	4
%SOM (Relative)	2.32	2.16	2.08	2.01
	(100)	(93)	(90)	(87)
Instability (Relative)	1.23	1.25	1.33	1.39
	(100)	(102)	(108)	(113)
% Closed area (crust formation)	82.97	88.39	90.62	91.93
(Relative)	(100)	(107)	(109)	(111)

[1]   SOM-content at the start = 2.08%

Object

1. 40 ton ha^{-1} farmyard manure every three years + green manure every three years
2. Input of all crop residues + green manure every three years
3. Green manure every three years
4. Blank, no supplementary input of organic material except roots and stubbles

water supply through improved infiltration, decreased loss via evaporation, better drainage of heavy-textured soils, resulting in an earlier and better workability of these soils, increased water holding capacity of light textured soils, more extensive and deeper root systems that make more moisture available for crop use and increased yields that result in more harvested crop per unit of water lost by evapotranspiration.

Influence on chemical soil fertility

Crop yields are commonly limited more by an inadequate supply of nitrogen than of any other element and hence this element occupies an unique metabolic position. SOM constitutes a nitrogen reservoir that is never completely empty but yet seldom full enough to provide enough of the element to permit maximum growth of potatoes. But the fact that SOM furnishes a regular supply of available nitrogen constitutes one of the main benefits derived from its presence, especially when mineral nitrogen, mostly present

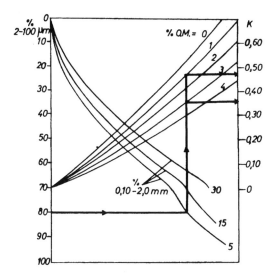

Figure 1   Nomogram: erodibility of soils (Wischmeier & Smith, 1978).

as NO_3^--N, is leached out during wet periods, and cannot fully recover plant's N-needs. As mineralisation goes on when plants are not growing, only a certain percentage of the mineralised nitrogen can be accounted for in modern, scientifically based nitrogen fertilisation advice systems, for potatoes, 75 to 80%.

Besides nitrogen, as the main plant nutrient, phosphorus and to a lesser extent potassium can be built up in the soil profile, creating a better chemical soil fertility.
SOM is an important source of phosphorus for plants, usually constituting 15-80% of the total phosphorus in soil (Allison, 1973). Organic phosphorus compounds in soil behave very much as do organic nitrogen substances with respect to decomposition. Plants can absorb some soluble organic phosphates, but most of the phosphorus undoubtedly enters the root as the primary or secondary orthophosphate ion.
Potatoes also need a lot of potassium and as such, organic matter, by means of its cation exchange complex does decrease loss by leaching and prevents inorganic fixation. Furthermore, plant residues that are high in potassium also serve as excellent, readily-available sources of this element. Since this nutrient

is not in organic form in plants in the sense that nitrogen and phosphorus are, much of it is available for crop use immediately after the plant residues are added to soil (Allison,1973).

Sources, composition and effects of organic materials

To maintain SOM at a certain level, fresh organic materials have to be incorporated into the soil. The most important ones are, besides animal manures, green manures, crop residues (straw from cereals, leaves and tops from sugar beets,...) and all kinds of secondary products (composts, sludges,...). In case of these secondary products, care has to be taken to the presence of heavy metals.

Animal manure

The composition of the diverse organic materials is different and even the 'same' organic material, e.g. slurry, changes from place to place and in time. Therefore, Table 2 only gives an 'average' composition of some animal manures and substantial differences can be found when analysing samples.

Though important towards SOM building (see further), total composition gives no idea about the amount of nutrients which is likely to be released from organic manures. In Table 3, the average available amounts of nutrients coming from animal manures are given, but also here, depending on type of manure (slurry, farmyard manure,...), time of application (autumn or spring essentially), ... large variations are to be expected.

Green manures and harvest residues

Other important sources of organic material are green manures and harvest residues. Table 4 gives the yield and composition of some green manures and harvest residues. Just as animal manures, they are also important towards SOM building (see further).

Table 2   *The average composition of animal manures (Bries et al., 1995).*

	Dry matter	Organic matter	Total N	P_2O_5	K_2O
Slurry (kg $(1000\ l)^{-1}$)					
Cow	95	60	4.4	1.8	5.4
Calf	20	15	3.0	1.3	3.0
Pig	85	60	6.5	4.0	5.2
Sow	60	40	4.0	3.7	3.6
Chicken	160	90	9.0	8.3	5.6
Farmyard manure (kg $(1000\ kg)^{-1}$)					
Cow	240	140	5.5	3.5	6.0
Pig	230	160	7.5	9.0	3.5
Chicken (wet)	320	230	12.5	15.0	11.0
Chicken (dry)	600	420	24.0	28.0	21.0
Broiler	580	460	26.0	19.0	20.6
Champost	400	200	7.0	6.0	9.0

As with animal manures, green manures and harvest residues are sources of nutrients, of which nitrogen is the most important one. When ploughing in harvest residues, N is mostly to be released through mineralisation within some weeks (Schrage & Scharpf, 1987) and leached out during the subsequent winter. Concerning green manures, depending on the time of incorporation, more or less nitrogen can be available towards the following crop. As potatoes are susceptible to a good structure, ploughing in before winter is preferred. Consequently, N is also likely to be released and depending on winter rainfall, be leached out. Nevertheless, when incorporated in spring, part of the N will already be released and subsequently measured prior to N-fertilisation, another part is likely to be mineralised during cultivation and therefore needs to be estimated.

Table 3   *Average amount of available nutrients coming from animal manures (Bries et al.,*
              1995).

	N	P_2O_5	K_2O
		Spring application (all soil types)	
Slurry (kg (1000 l)$^{-1}$)			
Cow	25	13	49
Calf	19	9	27
Pig	41	40	47
Sow	25	33	32
Chicken	63	58	50
Farmyard manure (kg (1000 kg)$^{-1}$)			
Cow	16	21	40
Pig	22	63	25
Chicken (wet)	68	105	99
Chicken (dry)	129	196	189
Broiler	142	133	185
Champost	18	36	58
		Autumn application (sandy soils)	
Slurry (kg (1000 l)$^{-1}$)			
Cow	9	13	14
Calf	5	9	8
Pig	17	40	13
Sow	10	33	9
Chicken	18	58	14
Farmyard manure (kg (1000 kg)$^{-1}$)			
Cow	9	21	18
Pig	12	63	11
Chicken (wet)	41	105	28
Chicken (dry)	82	196	53
Broiler	87	133	51
Champost	12	36	27
		Autumn application (heavy soils)	
Slurry (kg (1000 l)$^{-1}$)			
Cow	12	13	38
Calf	8	9	21
Pig	21	40	36
Sow	13	33	25
Chicken	26	58	39
Farmyard manure (kg (1000 kg)$^{-1}$)			
Cow	10	21	30
Pig	14	63	18
Chicken (wet)	48	105	77
Chicken (dry)	93	196	147
Broiler	100	133	144

Champost	13	36	45

Table 4   *Dry matter yield and composition of some green manures and harvest residues (Anonymous, 1980; Ninane et al., 1995).*

	Organic dry matter yield (kg ha^{-1})	Total N (% dry matter)
White clover	3500	2.95-3.15
Ray-grass (Italian)	4500-5000	1.65-1.76
Vetch	3000-4500	2.42-3.60
Mustard	3500-6500	1.63-2.82
Lucerne	4500-6500	3.0 - 3.2
Phacelia	3500-6000	1.44-1.85
Winter wheat		
Roots + stubble	3700	1.2
Straw	7000	0.6
Winter barley		
Roots + stubble	3500	1.2
Straw	6500	0.5
Sugar-beet		
Roots	500	1.5
Tops and leaves	5500	2.2 - 2.6

Mineralisation and humification

Besides nutrient supply through mineralisation, organic materials can contribute towards the building up of SOM through humification. Theoretically, the variation in time of the total amount of organic matter in the arable layer can be given by the following exponential equation:

$$mK_1 - BK_2 = \frac{dB}{dt}$$

whereby

m   :   annual input of organic material (kg ha^{-1})

K_1 : isohumic coefficient
mK_1 : quantity of organic matter formed each year (kg ha^{-1})
B   : amount of SOM in the arable layer
K_2 : mineralisation coefficient
BK_2 : quantity of SOM mineralised each year (kg ha^{-1})
t   : time in years

Accepting mK_1 is constant and K_2 is known, integration of the former equation gives:

$$B = \frac{mK_1}{K_2} - (\frac{mK_1}{K_2} - B_0).e^{-K_2 t}$$

whereby
B_0 : the original amount of SOM

Adaptation of this equation on measurements of B over time will allow to calculate a global K_1 (Table 5) and the evolution of B on the long term.

Besides the calculation of this humification coefficient, another approach, determining 'effective organic matter content', is in use. The effective organic matter content is the amount of organic matter which remains in the soil one year after incorporation (Table 6). This approach is more easy to use when calculating an organic matter balance.

Table 5   *Some humification coefficients.*

	K_1
Leaves	0.20
Roots	0.35
Green manures	0.25
Straw	0.30
Farmyard manure	0.40-0.50

Table 6   Effective organic matter content (kg ha⁻¹) (Anonymous, 1980 & 1989).

Winter-wheat (roots and stubble)	1100
Winter-wheat (straw)	2250
Winter-barley (roots and stubble)	1000
Winter-barley (straw)	1950
Oats	2470
Rye	2520
Potatoes	775
Tops and leaves of sugar-beet	960
Sugar beets (no leaves)	375
White clover	850
Ray-grass (Italian)	1250
Vetch	645
Lucerne	1700
Cow slurry (kg ton⁻¹)	30
Pig slurry (kg ton⁻¹)	25
Chicken slurry (kg ton⁻¹)	45
Cow manure (kg ton⁻¹)	70
Chicken manure (dry) (kg ton⁻¹)	200
Broiler manure (kg ton⁻¹)	225
Champost (kg ton⁻¹)	100

Organic matter balance

Each year part of SOM is broken down (mineralisation). A 1 hectare field, weighing 3500000 kg of soil and containing 1.4% SOC (2.4% SOM) contains about 85000 kg SOM. Having a mineralisation coefficient of 2% (moderate climate), this results in a yearly breakdown of ±1700 kg effective organic matter. To compensate for this annual loss, there is a need for 1700 kg effective organic matter or 5100 kg during a three year rotation. Out of the above mentioned numbers, an organic matter balance can be made up, calculating if enough organic matter is added. Table 7 shows the same rotation, but one with a bad and one with a good organic matter management.

Table 7   *Organic matter balances for the same rotation, using different sources of organic matter.*

Rotation (three years)	Effective organic matter (kg ha^{-1})	Annual SOM-breakdown (kg ha^{-1})
Winter-wheat (roots and stubble)	1100	1700
Sugar-beets (no leaves)	375	1700
Potatoes	775	1700
Total amount	2250	5100
	Shortage: 2850 kg	
Winter-wheat (roots and stubble)	1100	1700
Sugar beets (tops + leaves)	960	1700
Potatoes	775	1700
+ 40 ton cow manure	2800	
Total amount	5635	5100
	Surplus: 535 kg	

Soil organic matter level

The optimum organic matter content of the soil depends on the desired cultural effect and as such changes from field to field. The desired SOM content, from a physical point of view, is given by:

$$\frac{\% SOM}{\% Clay} * 100 > 7 \quad \text{(Vilain, 1996)}$$

However to compensate for a low mineral colloidal fraction, i.e. a soil constituting a low amount of clay, a high SOM is desired. This means that both heavy clay soils and light sandy soils need a high SOM content but due to different reasons.

The biggest inconvenient of a high SOM content is the fact that, due to an uncertain high N-mineralisation, a correct N-fertilisation advice is difficult to give.

Target values for different texture classes are given in Table 8.

Table 8   Soil Organic Carbon target values for 3 different texture classes.

Texture class	% Soil Organic Carbon
Sandy	1.1-1.4
Sandy loam - loam	1.0-1.3
Clay	1.3-1.6

Organic matter and potato production

The effect of amendments of fresh organic materials on tuber yield is not pronounced, at least in soils with a high mineralisation capacity. Table 9 shows the results of a field trial with amendments of different amounts of organic material combined with different amounts of mineral fertiliser (Bries et al., 1995). Besides quantity, also different quality parameters were investigated. Quality wasn't neither positively nor negatively influenced by the use of different organic

Table 9   Total tuber yield (relative) for different amendments of organic material and mineral fertiliser (Bries et al., 1995).

	1990		1991		1992	
	Mineral N (kg N ha^{-1})	Yield	Mineral N (kg N ha^{-1})	Yield	Mineral N (kg N ha^{-1})	Yield
Slurry[*]	0	118	0	121	0	121
Stable manure[**]	0	114	0	108	0	116
None	0	100	0	100	0	100
Slurry	241	117	150	112	196	113
Stable manure	241	120	150	97	196	123
None	232	120	186	110	196	123
100% (kg ha^{-1}) =		37333		36674		54900

[*]   Pig slurry: 55-60 ton ha^{-1}
[**]   Cow manure: 52-60 ton ha^{-1}

matter sources. Other field trials (Ampe, 1991) showed some negative influences when using slurry at too high amounts (83 ton ha^{-1}) just before planting.

On the contrary to above mentioned results, Polish research shows a beneficial effect both towards yield and the restriction of diseases (Table 10).
Moreover, farmyard manure composition (undecomposed straw content) or degree of decomposition of green manure at ploughing time can significantly affect potato yield and optimal N fertilisation rate by influencing soil mineral nitrogen availability next summer (Goffart et al., 1997).

Table 10   *The influence of fertilisation treatments on potato growth and tubers infection with Rhizoctonia solani.*

Parameter	Fertilisation treatments			
	PK	NPK	slurry	
			from cattle	from pigs
Intensity of tubers and sprouts infestation by Rhizoctonia [*]	2.3	2.1	0.3	0.4
The number of shoots per plant	2.6	2.6	3.6	3.5
Average height of shoots (cm)	28.0	26.9	43.2	44.1
Tuber yield t/ha	13.6	19.8	26.1	27.2

[*]   Due to 9-degree scale; 0 - no symptoms of disease; 9 - the most intensive infestation

Conclusions

The role and value of organic matter is diverse and depending on the desired effect, different actions can be taken.

SOM exerts its effect both towards physical soil conditions, through its action on aggregate stability and soil conservation, as to chemical soil fertility, through a regular supply of plant nutrients.
Fresh organic material can be subdivided into two groups: animal manures and green manures/harvest residues.

Keeping into account mineralisation and humification allows a sound organic matter management and therefore keeping the soil in good condition.

The beneficial effect of the addition of fresh organic material on potato yield depends on the soil fertility level. Too high amounts of slurry should be avoided for quality reasons.

3.3   Losses of mineral nitrogen from the plant-soil system in potato production

R. Postma, J.P. Goffart, P.A. Johnson, J. Salomez, M.A. Shepherd & R.E. Wheatley

Crop production has increased enormously in recent decades, and so, consequently have nutrient inputs. High inputs inevitably result in some leakage of the nutrients. Nitrogen can be lost from the plant-soil system by several processes, such as leaching, denitrification and volatilization.
High losses are generally undesirable as they are damaging to the environment and a waste of resource. In this subchapter, the nitrogen loss during and after the growth of potatoes will be discussed. An attempt will be made to answer questions such as: How much nitrogen is lost during the cultivation of potatoes in comparison to other crops? What are the main processes by which nitrogen is lost and what are the factors that control them? How can these processes be measured and under what conditions do they occur? What is the relative importance of different loss processes, and how can the losses of nitrogen be reduced?

Introduction

Initial information on the fate of nutrients within an agricultural system can be obtained by constructing a budget that accounts for inputs and outputs together with balances. Balances can be calculated over a range of different scales and complexities. In the simplest case, the N input to a field (in fertilizers and manure) and the N output from a field (in harvest products) are considered (Table 1). In some countries in Europe measures that are taken to reduce the losses to the environment, are based on such balances.

From Table 1 it can be seen that the surplus in the N balance is highest with sugar beet and lowest with spring wheat, and is greater with high levels of input than reduced. So, is the risk of nitrogen losses greatest with sugarbeet? This question is not easy to answer, because the type of N causing the surplus

Table 1   Simplified nitrogen balance for potatoes, sugarbeet and spring wheat at high and
          reduced levels of N input in the Netherlands (Vos, 1996). The amount of N is
          expressed in kg N ha^{-1}.

	potatoes		sugarbeet		s. wheat	
	reduced N	high N	reduced N	high N	reduced N	high N
N input mineral fertilizer	170	234	152	204	107	145
N output harvest product	181	204	87	100	120	128
N surplus	-11	+30	+65	+104	-13	+17

differs with the different crops. The nitrogen can be lost during the growing period or remain as organic N in the crop residues or as organic and mineral N in the soil. The amount of N in sugarbeet crop residues is relatively high, the averages for potatoes, sugarbeet and wheat are 35, 120 and 25 kg N per ha, respectively (Velthof et al., 1999). The N in these crop residues will gradually mineralize, and the rate and timing of this mineralization will affect losses: e.g. if the N is mineralized before winter, it may easily be lost, but if the N is mineralized after winter, it will be partly used by the following crop.

So, the total N balance for a field only gives a very broad indication of the risk of loss. In addition, conversions from organic to inorganic N in the soil and vice versa (mineralization and immobilization) are important in this risk assessment. The rate and timing of the mineralization will determine the amount of mineral N (NH_4^+ and NO_3^-) present in the soil. This mineral N may be lost by leaching, denitrification and/or volatilization. An example of the balance sheet of mineral N in the root zone of potatoes for one year is given in Table 2.

During the year (from March to March) the soil mineral N pool remained more or less the same, but nitrogen losses during the growth of potatoes may be substantial (Table 2).

Table 2   Nitrogen balance in a potato crop in Denmark over one year (from March to March; after Jensen et al., 1994).

Input of mineral N to the root zone	Amount of N, kg N ha^{-1}
Mineral N, including mineral N in organic manure and	115
Atmospheric deposition	18
Net mineralization from soil organic matter, crop residues and organic manure	99
Output of mineral N	
Total plant uptake	138
Nitrogen losses, i.e. NH$_3$ volatilization, denitrification and leaching	92
Unaccounted for	2

In general, it is important to know which process (volatilization, denitrification, or leaching) is responsible for the N loss as i) the impact on the environment differs between the loss processes and ii) this knowledge is required for the development of tools to reduce the losses. Therefore, some general principles and means of measuring and controlling the different loss processes are discussed.

Loss processes, controlling factors and means of determination

Leaching

N is leached from the rooting zone to deeper soil layers during periods with either a precipitation or irrigation surplus. In this way N is transported down the soil profile out of reach of the plant roots, and so is lost from the plant-soil system. N is mainly leached in the form of NO_3^-. Important factors affecting the amount of N that is lost by leaching include the amount of NO_3^- in the root zone, the precipitation surplus and rooting depth. Moreover, soil properties such as soil texture, soil structure and drainage properties affect N leaching.

In most cases the amount of leaching is determined by measuring [N] in the soil water at certain depth and calculating the amount of drainage water. Multiplying the two will result in the total amount of N leached.

$L = [N]_{drained} \times W_{drained}$, where

- L = amount of nitrogen leached (kg ha^{-1}),
- $[N]_{drained}$ = nitrogen concentration in the drain water (kg m^3),
- $W_{drained}$ = amount of water drained per hectare (m^3 ha^{-1}).

Denitrification

Denitrification is the microbiological reduction of NO_3^- or nitrite (NO_2^-) via several intermediate products into the gases nitrous oxide (N_2O) and molecular nitrogen (N_2).

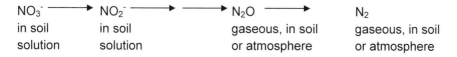

NO_3^-	NO_2^-	N_2O	N_2
in soil	in soil	gaseous, in soil	gaseous, in soil
solution	solution	or atmosphere	or atmosphere

Once NO_3^- in the soil has been converted into these gases, it is no longer available for plant uptake. Denitrification in soils takes place under anaerobic conditions (absence of oxygen; wet soils), and is also strongly dependent on the availability of organic compounds. NO_3^- is required for denitrification, but at $[NO_3^-]$ prevalent in most agricultural soils, denitrification rate is usually independent of $[NO_3^-]$. However, in microsites of agricultural soils, $[NO_3^-]$ may be limiting, so that the application of NO_3^- to soil will increase denitrification. The minimum temperature for the occurrence of denitrification in soil ranges from 2 to 10°C and the optimum temperature generally is about 30°C. So, it can occur, if slowly, throughout most of a northern European winter and may be particularly rapid in southern European and Mediterranean soils.

Actual denitrification under field conditions is difficult to measure. For that reason, denitrification was often estimated by the amount of N not accounted for in the N balance. Actual denitrification rates may be determined using the acetylene inhibition method with undisturbed soil cores (Aulakh et al., 1992). With this method the increase in the concentration of N_2O in a flux chamber is measured during a period of about 24 hours.

$D = (\Delta g/\Delta t * v)/ a$, where

- D   = denitrification rate $(g\ m^{-2}\ d^{-1})$
- g   = concentration of N_2O and/or N_2 $(g\ m^{-3})$
- t   = time (d)
- v   = air volume in flux chamber (m^3)
- a   = surface area of soil under consideration (m^2)

Volatilization

Volatilization of ammonia (NH_3) is another process by which gaseous N may be lost from the plant-soil system. The conversion of NH_4^+ into NH_3 takes place especially at high pH.

NH_4^+ ⇌	NH_4^+ ⇌	$NH_3 + H^+$ ⇌	NH_3 ⇌	NH_3
adsorbed	in soil solution	in soil solution	gaseous, in soil	gaseous in atmosphere

There is general agreement that most NH_3 emitted to the atmosphere in Europe originates from farm animals and their waste. Other sources such as fertilizer application, are minor in comparison (Ecetoc, 1994). The major part of NH_3 emissions from manure application comes from manure left on the soil surface after spreading. Important factors controlling NH_3 volatilization from soil are the ammonia concentration and pH of the soil.

Ammonia volatilization in the field may be measured with a micrometeorological mass balance method (Denmead, 1983). With this method, the NH_3 flux from the soil is calculated by a gradient of NH_3 concentrations in the air above the surface together with wind speed.

$A = -K * \Delta g/\Delta z$, where

- A   = NH_3 volatilization $(g\ m^{-2}\ d^{-1})$,
- K   = eddy diffusivity or transport coefficient of the gas $(m^2\ d^{-1})$,
- g   = concentration of NH_3 $(g\ m^{-3})$,
- z   = height above the surface (m).

Nitrification

Nitrification, the conversion of NH_4^+ into NO_3^-, is not a loss as such. However, it is of indirect importance for the loss of mineral nitrogen from the plant-root system. The conversion of NH_4^+ to NO_3^- results in a product that behaves differently in the soil: NO_3^- is more susceptible to leaching as NH_4^+ is adsorbed onto the clay and organic matter particles, whereas NO_3^- remains in the soil solution.

Moreover, NO_3^- may be lost from the soil solution by denitrification. Nitrification is a microbiological process, which takes place under aerobic conditions. The process can be subdivided into two steps, i.e. the conversion of NH_4^+ into NO_2^-, and the conversion of NO_2^- into NO_3^-. In general, the conversion of NO_2^- into NO_3^- proceeds much faster than the conversion of NH_4^+ into NO_2^-, thus preventing the accumulation of (toxic) NO_2^- in soil. Nitrification proceeds if the supplies of ammonium (NH_4^+) and oxygen (O_2) are sufficiently high. O_2 is limited at high soil moisture contents, high soil temperatures and high contents of oxidisable organic matter. Moreover, the pH of the soil should be within the range of 4 and 8, and the optimum temperature for nitrification appears to vary between 20 and 40°C. No nitrification takes place if the temperature is below 5°C.

Denitrification and nitrification may both lead to the emission of nitrous oxide (N_2O) from the soil surface. This is of concern from an environmental point of view, because N_2O is a greenhouse gas and it has a destructive effect on the ozone layer. The effects of denitrification and nitrification on N_2O emission are not dealt with in this chapter, because we focus on the total amount of N that gets lost by the different processes. However, measures taken to reduce denitrification will reduce the emission of N_2O as well.

Losses during the growing period

Several factors contribute to the loss of N during the growing period. Mineral N content, soil moisture content and drainage are important factors affecting leaching and denitrification. The amount of mineral N in the soil during the growing period is determined by that present before planting, the amount and time of fertilizer application, and the N mineralization from soil organic and fresh organic materials. Crop uptake of N is also relevant. Soil moisture content and drainage are affected by the weather, soil texture, ground water level, irrigation and uptake by the crop.

Powlson (1997) indicated that a combination of high N content in the soil and high precipitation, such as rainfall during the first three weeks after fertilizer application, was a major factor in describing nitrogen losses in winter wheat. MacDonald et al. (1997) found a comparable effect for potatoes in the United Kingdom. The proportion of the fertilizer N lost after application to potatoes was between 10 and 20%.

Not many results exist of direct measurements of NO_3^- leaching during the growth of potatoes. Shepherd (1992) showed that a substantial amount of N (40-80 kg N per ha) can be lost, when leaching occurs after fertilizer application and mineral N contents are lower if drainage is high in the period between fertilizer application and tuber initiation (Table 3). It was suggested that leaching was mainly responsible for the decrease in mineral N in the sandy soil, but denitrification might also have played a role. The leaching can follow rainfall or irrigation, if they take the level of soil moisture above field capacity.

Measurements of actual denitrification have been carried out with several crops. Denitrification rates vary between 0 and 100 kg N per ha per year, and peaks in denitrification rate are found after the application of N fertilizer, during wet periods, or after the supply of easily decomposable organic C (Aulakh et al., 1992). Because denitrification may be high during short periods ('peaks'), reliable estimates of the annual denitrification loss may require daily measurements (Aulakh et al., 1992).

Table 3   *Soil mineral N and crop N (kg N ha^{-1}) after drainage between planting of potatoes and tuber initiation. Samples were taken late May or early June. (After Shepherd, 1992).*

Drainage, mm	depth, cm				crop N
	0-30	30-60	60-90	0-90	
17	102	42	36	180	32
74	60	36	27	123	33
123	40	22	19	81	28

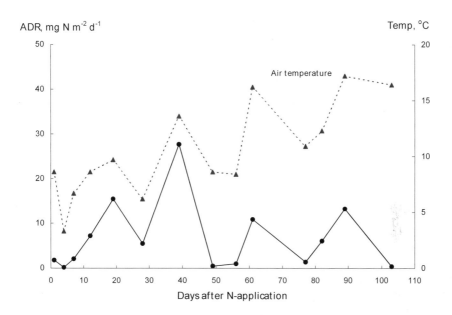

Figure 1   *Rates of actual denitrification (ADR) in the top 20 cm of a non-fertilized potato field on a sandy loam soil in the Netherlands (Postma and van Loon, 1996).*

Studies on actual denitrification during the growth of potatoes are limited. An example of the monitoring of actual denitrification during the growth of potatoes is given in Figure 1. The application of 200 kg N as calcium ammonium nitrate (CAN) did not significantly increase denitrification, indicating that the process was not limited by the NO_3^- supply, but by another factor. Probably, the O_2 concentration was too high to enable a high denitrification rate. Denitrification rate fluctuated with temperature. The total nitrogen loss during the first 100 days after planting amounted to 6 to 8 kg N per ha for both the fertilized and unfertilized situation.

It is generally known that the spatial variation of denitrification and the emission of N_2O within a field is very high. Coefficients of variation are often higher than 100 % (Aulakh et al., 1992). In potato fields, the spatial variation in denitrification and nitrous oxide emission (McTaggart and Smith, 1996) is partly caused by the differences between furrows and ridges. In a sandy loam soil in the Netherlands, the average denitrification rate in furrows was 5.0 mg N per m^2

per day and in ridges it was 0.4 mg N per m^2 per day (Postma and van Loon, 1996).

Often, results of actual denitrification rates are rather low. However, potential denitrification rates -denitrification rates estimated after removal of O_2 and the addition of water and NO_3^-- are much higher and may be several kg N per ha per day (Wheatley et al., 1991). They found that the potential denitrification rate fluctuated during the growing period of potatoes, due to variations in the supply of easily utilisable C.

It is questionable whether the optimal conditions for denitrification ever occur in the field and, if so, for how long they persist. This is most likely to occur during periods of high rainfall, or irrigation and high temperatures immediately after fertilizer application.

The risk of NH_3 volatilization is especially high following the application of animal manure to the soil. The largest part of the volatilization occurring immediately after application and may amount to 50% of the NH_4^+ applied in the manure (Fig. 2). In Figure 2 it is shown that the volatilization is greatly reduced when the slurry is injected into the soil.

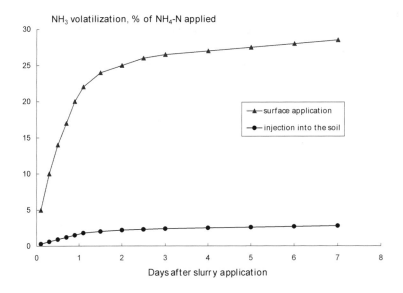

Figure 2   Pattern of NH_3 volatilization after the application of cattle slurry to a light textured clay soil without plants in Germany (after Dosch and Gutser, 1996).

NH_3 volatilization may also occur after the application of mineral fertilizers which contain NH_4^+ or urea. NH_3 emission following fertilizer application is caused by chemical reactions with soil components, after the fertilizer has been applied. Soils differ in their reactivity and in their ability to provide conditions conducive for NH_3 emissions. Emissions tend to be highest from calcareous rather than acid soil types.

The risk of NH_3 volatilization varies with fertilizer type. In general, the risks are the highest after the application or urea, which has been studied extensively. The loss of nitrogen by NH_3 volatilization after urea application is generally in the range of 10 to 25%. Emissions from other NH_4^+ containing fertilizers are markedly lower than those of urea, less than 10% of the N applied, with the possible exception of ammonium sulphate and diammonium phosphate on calcareous soils.

According to an estimate of the total NH_3 volatilization in Europe (Ecetoc, 1994) half of the NH_3 volatilization from fertilizer application is associated with the use of urea, while urea accounted only for 18% of the N application in Europe in 1990. The main N fertilizers used in Europe are ammonium nitrates. However, in some regions of southern Europe urea is the dominant N source.

Some NH_3 volatilization may also originate from plants, especially at high N fertilization levels. Estimations are 1 kg N per ha per year from cereals in Denmark and 10 kg N per ha per year in the South of the UK (Ecetoc ,1994). Decaying vegetation gives off NH_3, especially if it is rich in N. This amount is estimated to be 1.5 to 6 kg N per ha per year.

Losses after the growing period

The risk of losses of nitrogen after crop harvest is often higher than during the growing period. In the introduction it has already been mentioned that some N will remain in the soil in the form of mineral N after harvest and some N will remain on top of the soil as crop residues.

Because of the precipitation surplus in autumn and winter in large parts of Europe, the mineral N remaining in the soil after harvest is liable to leaching and denitrification. The amount of residual N after potatoes is greater than after other crops (e.g. Shepherd and Lord, 1996). Shepherd and Lord (1996) compared the amount of mineral N after harvest of potatoes, sugarbeet and winter wheat. Mineral N levels after potatoes were the highest (60 kg N per ha after the application of 220 kg N) and lowest after sugarbeet (20 kg N per ha after the application of 125 kg N).

The amount of mineral N that remains in the soil at harvest is influenced by the amount of fertilizer applied and the input of organic materials. The relationship between the amount of N applied with fertilizers and the residual amount of mineral N at harvest is not linear. For low levels of application it remains constant, but then rapidly increases (Fig. 3).
The N that remains on top of the soil in the form of crop residues, will partly mineralize during autumn and winter. Total N mineralization from crop residues depends on the total amount in the residues, the decomposability of the material, the C:N ratio and the time of incorporation of the material.

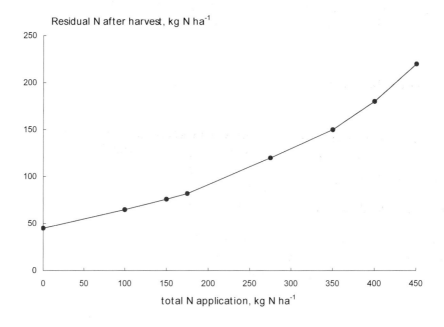

Figure 3   *Relationship between the total N applied in fertilizer and the residual N in the soil at*
            harvest on a sandy soil in the Netherlands (Neeteson, 1992).

Average amounts of N in crop residues are 35, 120 and 25 kg N per ha for potatoes, sugarbeet and wheat. Moreover, the decomposability of sugarbeet tops is relatively high and the C:N ratio relatively low, (15:1; for wheat and potatoes it is 60:1 and 35:1, respectively). This will result in rapid N mineralization from sugarbeet tops, which will increase the risks of losses of nitrogen from this crop during winter.

Of course mineralization of soil organic matter will also go on after harvest of the crop. Especially fresh organic matter, originating from animal manure and crop residues will be mineralized relatively easily.

What happens to the N remaining in the soil at harvest or that converted into mineral N during autumn and winter? Probably, most of it will be lost, but how? What is the relative importance of different loss mechanisms? Shepherd and Lord (1996) found fluctuations in leaching losses after harvest of potatoes from 40 to 140 kg N per ha, but they did not study denitrification and NH_3 volatilization at the same time.

Factors that influence the relative importance of different loss mechanisms

Addiscot and Powlson (1992) estimated the partitioning between leaching and denitrification in 13 experiments with winter wheat. The total nitrogen loss varied between 1 and 35% of the total N applied (150 kg N per ha) and in 9 of the 13 experiments the estimates for denitrification were higher than for leaching. The average total loss was 15.7%, of which denitrification contributed 10.0% and leaching 5.7%.

In general a qualitative indication of the partitioning of losses of nitrogen between leaching, denitrification and NH_3 volatilization may be obtained from an evaluation of conditions, such as soil type, weather conditions and management practices.

Soil type

As has been stated before, the risk of NO_3^- leaching and denitrification will be high if the NO_3^- content in the soil is high. In soils with a high mineralization potential, NO_3^- may be formed during periods the crop does not require N, and in these soils there is a high risk of both NO_3^- leaching and denitrification. The risk of NO_3^- leaching will be higher on sandy soil, as the water holding capacity of sandy soils is limited, and the risk of denitrificaton will be relatively higher in clay soils and soils with bad drainage or a high groundwater table. The presence of easily decomposable organic matter will also favor denitrification.

NH_3 volatilization from NH_4^+ or urea fertilizers will be especially high if soil pH is high and the clay fraction is low. With respect to the surface application of animal manure, the infiltration properties of the soil are also of importance, because the NH_3 volatilization from animal manure starts as soon as it is exposed to air. This exposure is reduced if the manure has been infiltrated into the soil.

Weather conditions

The risks of NO_3^- leaching are the highest during winter, because of the precipitation surplus in most regions of Europe. There is also a risk of NO_3^- leaching after fertilizer application in early spring, when the crop is not yet fully developed and evapotranspiration is still low.

The risk of denitrification losses are highest during spring and autumn, because the weather may be both wet and warm. Denitrification is affected indirectly by a combination of precipitation surplus, affecting soil moisture content and so anaerobic conditions. The optimum temperature for denitrification is about 30°C.

The risk of NH_3 volatilization is highest during warm and dry periods. Because NH_3 volatilization is reduced by precipitation after the application of NH_4^+ containing material to the soil, the time between manure or fertilizer application and the occurrence of precipitation events, is important. If this period is short, NH_3 volatilization will be limited. But if the period between the application of manure or fertilizer and precipitation is long, NH_3 volatilization may be relatively high. Moreover, NH_3 volatilization will increase with increasing temperature and windspeed.

Management practices

Irrigation is an important management practice that can increase losses of nitrogen. Excessive irrigation may increase N leaching by increased drainage through the soil (Shepherd, 1992), but it may also stimulate denitrification (Aulakh et al., 1992).

Ploughing may stimulate N mineralization (NH_4^+ formation) and nitrification (conversion of NH_4^+ into NO_3^-) by its positive effect on porosity and O_2 status.

The NO_3^- that has been formed in this way, may subsequently be leached or denitrified. However, ploughing will generally reduce denitrification by the positive effect on aeration (Aulakh et al., 1992).

Several factors affect the risk of losses of nitrogen after the application of fertilizers or manure. High N applications will generally lead to high losses of nitrogen and the time of application may also affect the risks. Application during periods with a high precipitation surplus, may lead to high losses of nitrogen by leaching, denitrification and volatilization. The material used as a fertilizer also affects losses. The use of animal manure or fertilizers which contain urea or NH_4^+ may lead to relatively high losses of NH_3 by volatilization, but losses by leaching and denitrification will be relatively low. Finally, application methods will have consequences for the level of nitrogen loss. E.g. the surface application of animal manure will lead to a higher NH_3 volatilization than after incorporation into the soil.

Growing cover crops to reduce nitrogen losses after harvest

We have already mentioned that the risk of losses of nitrogen is especially high after potatoes. This is due to relatively high amounts of mineral N that are left in the soil after harvest and the mineralization of soil organic matter and crop residues that may go on after harvest.

To reduce nitrogen losses after the harvest of the main crop, e.g. potatoes, catch or cover crops may be grown. Nitrogen loss will then be reduced by uptake from the soil into the new plants.

Vos and van der Putten (1997) studied the growth and N uptake of three catch crops in relation to sowing date in the Netherlands. They compared sowing at the end of August, mid September and the beginning of October. They concluded that the sowing date of catch crops is of great importance. If the crop is sown too late, growth will be insufficient and N uptake will be insignificant. They suggested to sow a catch crop before mid September in the Netherlands. In practice, this will only be possible after early potatoes.

They summarized the relationship between potential N uptake and sowing date as:

$Y = 960 - 3.4 \, X$, where

-   X is the day number in the year of the sowing date and
-   Y is the cumulated N uptake in kg N per ha (Vos and Van der Putten, 1997).

So, at day number 230 the N uptake is 178 kg N per ha and at day number 285 it equals zero.

Goffart et al. (1992) studied the effect of sowing different catch crops in July and August on mineral N contents during autumn and winter. The catch crops were ploughed in December and they reduced the mineral N content in the soil at that time with 40 to 80%. In spring, part of the residues of the catch crops were mineralized and the mineral N contents were relatively high in the top soil, as compared to a bare fallow.

Methods to reduce nitrogen losses during the growing period

Management measures to reduce nitrogen losses during and after the growing period should be focused on the minimisation of mineral N contents at times the crop does not need N to be taken up. So, a proper N management may be defined as the balancing of the supply of mineral N in the soil and the demand of N by the crop. If the supply and demand of N are in balance, excessive N contents in the soil will be prevented and the losses will be minimized.

With respect to the supply of N to the soil by fertilizers and/or animal manures, the following guidelines may be given (here we also refer to chapter 5.1):

-   the application of the optimum amount of N.
    As has been mentioned before, the application of high amounts of N will lead to high risks of nitrogen losses. For that purpose fertilizer recommendations are available. Generally, those recommendations make use of soil sampling, taking into account mineral N contents in the soil in spring and, eventually, the amount of N that may be mineralized in the course of the growing period.
-   apply mineral fertilizers and/or animal manure at the right time.
    Sometimes, animal manure or slurry is applied before winter. However, because this leads to high risks of leaching and denitrification, the application of animal manure in spring is preferred. Moreover, the period

between application of animal manure or mineral fertilizers and planting should be kept as short as possible, because the risk of leaching losses and denitrification is high in early spring.

All mineral N can be supplied as a basal application before planting, but it is also possible to split the N application into several dressings that are supplied during the growing period. In general, the risk of nitrogen losses will be lower if the N is split into several dressings. Furthermore, splitting the N into several dressings has the advantage that there are better possibilities to adjust the supply by the soil to the demand by the crop. Via intermediate determinations of the amount of mineral N in the soil or the N status of the crop, the amount of the seasonal N application can be optimised.

- make use of the most appropriate fertilizing substance.
  As has been mentioned before, the risk of ammonia volatilization from animal manures is high. This will take place before (from the stables, etc) and after application to the soil. Possibilities to reduce the ammonia volatilization from animal manures after application to the soil are:
  - dilution of the slurry with water
  - acidification of the slurry,
  - incorporation of the slurry into the soil (see following paragraph).
  Ammonia volatilization may take place as well after the application of NH_4^+ or urea containing fertilizers to the soil. Incorporation into the soil is also a possibility to reduce the ammonia volatilization from these mineral fertilizers. With nitrate containing fertilizers, there will be no risk of NH_3 volatilization. However, the risk of leaching and denitrification is relatively high with the use of NO_3^- containing fertilizers. After nitrification of NH_4^+ from animal manure and NH_4^+ or urea containing fertilizers, there is also a risk of leaching and denitrification. These risks may be reduced by the combined use with nitrification inhibitors.

- apply the mineral fertilizer and/or animal manure in the proper way.
  Animal manure can whether be applied on top of the soil or incorporated into the soil (see chapter 3.2). The incorporation into the soil will decrease ammonia volatilization and in several countries (e.g. the Netherlands) it is obliged to apply animal manure in that way.
  Mineral fertilizers may be broadcast, placed in the planting row, or foliar applied. In most countries the biggest part of mineral N is broadcast before planting and sometimes part of the mineral N is broadcast during the

growing period. In research, the possibility to increase the efficiency of N uptake by potatoes by placement of N fertilizer close to the planting tuber is studied. The objective is to increase the availability of N for uptake by the plant roots, especially during early growth stages. Moreover, the local application of NH_4^+ containing fertilizers may inhibit nitrification, thus keeping the N in a relatively immobile form. However, the results of this researches are confounding, until now. A special form of the application of mineral fertilizer during the growing period is foliar fertilization with liquid fertilizers. The idea is that N can be taken up directly by the leaves, thus acting very rapidly. However, the amount of N that can be taken up in this way is limited.

Excessive irrigation will increase nitrogen losses by leaching and denitrification. For that reason, a proper irrigation management is of great importance for the minimization of nitrogen losses. It is important to irrigate crops on drought-prone soils, because well-grown crops will use N more efficiently and leave less at harvest, thus reducing the risk of N leaching during winter.

Conditions with high risks of nitrogen losses and possible measures to reduce the losses are summarized in Table 4.

Table 4   Conditions with high risks of nitrogen losses and possible measures to reduce them.

Conditions with high risk of N losses			possible measures to
soil factors	weather conditions	management practices	reduce N losses
Nitrate leaching			
low water holding capacity, which is found in soils with low clay and organic matter contents	high precipitation surplus, which is especially the case durig winter, when evapotranspiration is low	leaving crop residues with high N contents at the field	apply the optimum amount of N
		ploughing before winter	apply animal manure in spring
high nitrate contents at times with high risk of leaching, which is found in soils with a high potential for N mineralization		applying high amounts of N containing fertilizing substances at times the crop does not require N	reduce the time between application and planting
			split the N dressing
		excessive irrigation	use NH_4^+ containing fertilizers with nitrification inhibitors
			do not plough before winter
			grow catch crops after harvest of main crop
denitrification			
high soil moisture contents, which are often found with: - high clay contents - bad drainage - high groundwater	high precipitation surplus, which is especially the case durig winter, when evapotranspiration is low	same as for nitrate leaching	same as for nitrate leaching
high nitrate contents	high temperatures		
high contents of easily decomposable organic matter			
ammonia volatilization			
low clay content	low precipitation frequency	use of animal manure and/or urea and/or ammonia containing mineral fertilizers	incorporate fertilizing substance into the soil
high pH			
low infiltration capacity	high temperature		remove crop residues with high N contents
		application at the time the soil surface is not covered	use nitrate containing fertilizers
		crop residues with high N contents	

Conclusions

The most important conclusions from this chapter are:
- Nitrogen losses associated with the growth of potatoes strongly depend on the conditions. In extreme cases the losses may amount to about 100 kg N per ha.
- Part of the N will be lost during the growing period and, probably the largest part, will be lost after the growing period.
- The relative importance of different loss mechanisms depends on the conditions, such as soil type, weather, fertilizer or manure used and way of application manures or fertilizers.
- Measures that can be taken to minimize the losses associated with the growth of potatoes are:
    - grow a catch crop after the harvest of potatoes, but (for western Europe) do not sow it after mid September,
    - make use of fertilizer recommendations, which take into account soil mineral N in spring and the expected mineralization during the growing period,
    - apply animal manure in spring and not in the preceding autumn or winter,
    - do not apply all the required N at planting, but withhold part until tuber initiation,
    - if animal manure is used, incorporate it into the soil.

3.4   Sampling and analysis of soils for soil nitrogen

P.A. Johnson, R. Postma, J.P. Goffart & J. Salomez

'Any analysis is only as good as the sample on which it is done'. How should soil samples for soil N be taken? What depth of soil should be sampled and what spatial arrangement of sampling points should be used? How should the samples be cared for whilst being moved from the field or experimental plot to the laboratory? Once in the laboratory what analysis technique should be used?

Introduction

Soil analysis for pH, phosphate potash and magnesium has been in use for many years, it is only more recently that soil analysis for nitrogen has become popular. Total soil N is not a good predictor of crop N requirement and there is a need to develop decision support systems (Chapter 3.5) which could use the results of testing for nitrate and ammonium N in the soil as a predictive tool. For various reasons explained elsewhere in this book both nitrate and ammonia are ephemeral when compared to the other major nutrients required by the potato crop. Because of variation of Mineral nitrogen within fields (Dampney et al., 1997) prior to the potato crop, after the crop and during cropping, it is essential that samples are taken carefully to fully represent the population being sampled. Precision farming (Chapter 6.2) takes this concept further. Samples must be taken to the correct depths for the predictive models being used, kept clear of contamination and correctly handled between field and laboratory otherwise the results of any analysis will be worthless. The analysis technique chosen will depend on the eventual use of the results.

This chapter aims to provide a robust sampling regime covering sampling methods, depths, patterns and sample handling prior to analysis for both field and experimental use and to discuss the analysis methods which may be appropriate.

Sampling Methods

Accuracy of sampling is vital no matter what the determination is to be. Sampling in this context is not just the taking of the sample in the field but also the sub-sampling which is done to reduce the sample size to that with which the analyst can cope.

Sampling for Total Nitrogen

Sampling for total nitrogen is normally only done on the topsoil to ploughing depth and is therefore physically undemanding. It is rare for samples to be taken for total nitrogen alone in consultancy though some prediction techniques use a total nitrogen figure determined on the 0-30 cm sample taken for Soil Mineral Nitrogen (nitrate and ammonium nitrogen) as a guide to potential mineralisation.

On a field basis a minimum of 15 (numbers vary from country to country) cores should be taken, though this number can be reduced on experimental plots, in an area not exceeding 10 ha which has been shown by prior examination to be of an even soil type and with little variation in history of cropping, fertiliser and organic manure use. Samples will normally be taken to 25 cm (ploughing depth) though there is merit in sampling only to 15 cm to reduce the risk of contamination with sub-soil material.

Sampling for Soil Mineral Nitrogen

The sampling process aims to reduce the chance of the later analytical procedures giving a result which is not representative of the area of land being characterised. It must also take into account the rooting depth of the crop to be or being grown.

A description of procedures used by ADAS in the UK is given below with additional comments added where the methods vary from a typical commercial organisation in the Netherlands.

For research work soil samples for Mineral nitrogen are normally taken to 90 cm depth in 3 increments of 30 cm. Exact details of the depths required will be given in the experimental protocol.

For consultancy sampling is normally carried out to a depth of 60 cm, though this will vary with the requirements of the predictive model being used.

A number of different auger types are available for Mineral nitrogen sampling as well as various vehicle mounted corers. It is important in taking the samples that there is no cross contamination between different depths.

Wherever possible samples should be analysed 'fresh' however in many situations this is not possible and therefore the samples submitted to the laboratory have to be deep frozen before being sent to the laboratory for extraction and analysis. Freezing the sample stops mineralisation which will invalidate the results. Large samples (over 1 kg) can generate a storage problem. Many laboratories require a minimum of 200 g of fresh soil and accurate sub-sampling in the laboratory is difficult, especially with heavy, cohesive soils. For these reasons, a sample from the field of around 500 g is preferable with an absolute maximum of 1 kg. Sub-samples resulting from the reduction of the original sample, can be retained as back-up samples.

Pre sampling preparation

Prepare enough bags for the number of plots/fields and the depths to be sampled. Label bags with waterproof ink and place an additional label within each bag (this label should be contained in a waterproof small polythene bag).

Always check that the area to be sampled has no contrasts, particularly in soil type, and has been managed and fertilised as one block, areas of contrasting history within one field should be sampled separately.

It is vital that there is no cross contamination between samples from different depths and therefore it is recommended that the top 2 cm of the sample, which may be contaminated with soil from the surface layers, is discarded.

Samples should not be taken within at least 6 weeks of the application of fertilisers or organic manures, though this period may be reduced post planting provided that this decision is documented in the report.

Sampling guidelines - consultancy

Take at least 10 cores per sample at the depths required by the model to be used for the prediction of N requirement, in the UK this would normally be 0-30 and 30-60 cm. The Dutch procedure suggests 0-30 cm on sandy and peaty soils (with 30 sampling points) and 0-30 and 30-60 cm on clayey soils (with 15 sampling points). The area sampled in the UK would not normally exceed 10 ha but in the Netherlands there is a recommended maximum area of 2 ha. Both countries recommend that samples are taken using a zig-zag or 'W' pattern in the field.

Samples will normally be taken in the spring for consultancy work after the majority of overwinter leaching has occurred.

Sampling guidelines - research

a) Pre-crop: On a field prior to an experiment, sample a minimum of 10 points, the area samples should not exceed 1 ha, avoiding atypical areas such as headlands, tramlines and wheelings sites of manure heaps, the area around drinking troughs and large trees, etc. More detailed sampling should be considered in many situations.

    When sampling plots prior to planting take at least 6 cores per sample at the sampling depths indicated in the protocol (usually 0-30, 30- 60 and 60-90 cm), select a minimum of 6 points avoiding a 0.5 m wide perimeter strip within the plot, and tramlines and wheelings. Where sequential sampling will take place it is necessary to record where samples are taken or use discrete plot areas for each sampling.

    Decide on the appropriate sampling pattern to use, this should normally be either a 'W' or grid pattern. Sampling points should be equally spaced throughout the area.

b) During cropping: Because potatoes are normally grown in ridges or well formed beds rather that on the flat like other crops sampling during the growing season is more complicated. Two methods are used. The first suggests that the samples are taken halfway between the top of the ridge and the furrow bottom and the second that ridges are flattened before sampling. The procedure for selecting sampling points described in Chapter 2.4 for plant sampling should be used.

c) Post cropping: Sampling should be done in the same manner as pre-planting and should where possible take into account the heterogeneity caused by the flattening out of the ridges in which the crop was grown.

Storage of samples

It is advisable to keep back-up samples of soil samples from experiments or consultancy services. To do this samples should be sub-sampled following return to base and back-up samples, labelled with the same details as on the original samples, frozen and retained until results are returned and verified.

Keep samples in a cool box while sampling and in transit.

On return from the field analyse the samples at once or place the samples in a freezer and store at -15°C or below before delivery to the laboratories in a cool box to ensure they remain frozen.

Spatial variation in soil nitrogen

Even a well taken sample will only give an average figure for the area sampled. Because of subtle changes in soil type and variations in residues from previous crops and manure applications there can be quite large variations in mineral nitrogen levels within a field. Within field variation for two fields in England (Dampney et al., 1997) are shown in Table 11. With N recommendation systems for potatoes commonly using soil mineral N (to 60 cm) values the ranges shown could have dramatic effects on N recommendations. Field A had grown winter wheat and Field B peas. Using a simplistic model the nitrogen requirement of a potato crop grown in field A could range from 180 to 265 kg/ha and for field B from 40 to 220 kg/ha N. Standard tables for the N requirement of the potato crop based on previous cropping would give reasonable recommendations for field A but would over-fertilised much of field B resulting in late maturity, a smaller yield of large potatoes and an increased level of residual mineral nitrogen at risk of leaching post harvest of the crop. The spatial variation in Mineral nitrogen

Table 1   Summary details of Mineral nitrogen and Total Nitrogen analysis.

Site	Analysis	Depth (cm)	Range		Median	CV%
Field A	Mineral nitrogen (kg/ha)	0-30	19.2	-50.8	31.0	23.3
		30-60	16.1	-69.6	28.8	27.4
	Total N (%)	0-30	0.06	-0.24	0.19	20.7
Field B	Mineral nitrogen (kg/ha)	0-30	24.2	-85.2	42.6	25.2
		30-60	52.4	-175.2	76.8	23.8
	Total N (%)	0-30	0.06	-0.23	0.18	10.4

for both fields is shown in Fig. 1 below. Precision farming techniques (Chapter 6.2) aim to take this variation into account when planning N fertiliser application rates.

Within the growing crop the spatial variation in Mineral nitrogen will in part depend on the method of application of N fertiliser as shown by the diagrams below (Hofman et al., 1993). which clearly show the effect of banding fertiliser N application. Where samples are to be taken from experimental treatments great care must be taken to ensure that the results clearly represent the potential variation in soil nitrogen concentrations.

Variation following the potato crop is also important as mineral nitrogen tests are more likely to be taken in consultancy in situations where residual mineral nitrogen is expected to be high rather than the expected lower levels prior to the

Field A   Field B

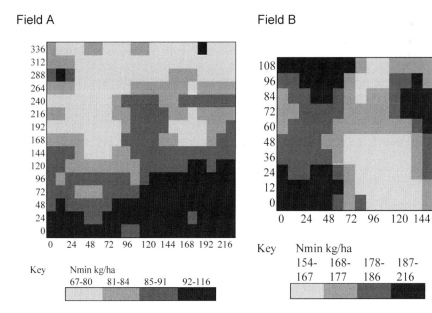

Figure 1   Spatial variation in Soil Mineral Nitrogen (kg/ha) in two English fields (Dampney et
             al., 1997).

             Table 1 and Figure 1 reproduced by permission of the authors and CIEC Editorial
             Board)

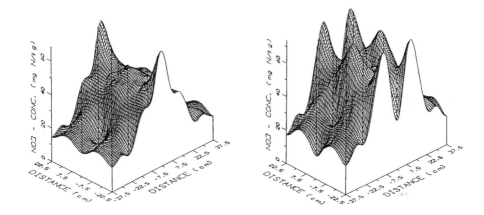

Figure 2   *Nitrate-N concentrations in soil following broadcasting (left) and placement of N fertiliser (right). (Reproduced with kind permission from Kluwer Academic Publishers.)*

potato crop which will commonly follow cereals which tend to leave low levels. Inaccurate application of N to the potato crop (probably overlapping) can influence the levels of residual soil mineral N (Van Meirvenne & Hofman, 1989), however by spring there was far less variation probably as the result of leaching.

Analysis Methods

A number of different aspects of soil N may be of importance to the farmer and the research scientist. Different centres may use different analysis techniques for the same 'determinant', some for historical reasons and some because they have found that the particular technique has been shown to be relevant for their local soils and cropping patterns.

Total Nitrogen

The Kjeldahl and Dumas techniques (as well as their modified versions) are accepted as being the standard methods of analysis of soil for total nitrogen, as they are for total plant nitrogen.

Available nitrogen

Available nitrogen is taken to be the sum of the nitrate and ammonium nitrogen extracted from the soil, this does not include any potentially available nitrogen from mineralisation during the growing season. A number of possible extractants with associated analytical techniques are shown in Table 2. Methods 3, 4 and 5 give comparable results for nitrate concentrations but the use of $CaCl_2$ as an extractant is less reliable for ammonium. It is likely that sample handling and treatment between field and analysis has a greater effect on the errors of determination than the actual analytical method. The need for rapid analysis of fresh samples or storage of samples in a frozen state has been referred to in the section on sampling. There are suggestions that the speed of thawing the sample before analysis may have an effect on Mineral nitrogen levels determined and some laboratories prefer to thaw the sample in a refrigerator rather than at ambient temperatures. Errors in the

Table 2   *Methods of analysis for available and potentially available nitrogen.*

Method	Suitable for Mineral N	Suitable for Potential available N
1. Kjeldahl [N]		yes
2. Dumas (including variants)		yes
3. $CaCl_2$ extraction + auto-analyser, dried soil	yes	
4. KCl extraction + auto-analyser, fresh soil	yes	
5. 1% $KAl(SO_4)_2$ + auto-analyser	yes	
6. KCl extraction (hot/cold), dried soil	yes	yes
7. KCl extraction (hot/cold), fresh soil	yes	yes
8. Electro Ultra Filtration	yes	
9. Cold Water Extraction	yes	

measurement of Mineral nitrogen following rapid thawing can be high in samples with low levels (e.g. less than 2 mg N/kg) but these soils are severely N limited anyway and the error is of little consequence in an advisory situation. In arable soils the error at greater concentrations is acceptable, but following grassland there are problems with the use of frozen samples and it is recommended that the samples are analysed fresh. The method of analysis used must always be quoted.

There is little doubt that drying samples at temperatures above ambient can increase the levels of ammonium recorded on analysis but has little effect on nitrate.

Potentially available nitrogen

Estimating the amount of N which may become available for use by the potato crop following planting, or after pre-planting sampling for mineral nitrogen, is more difficult. Predictive models such as WELLN (Rahn et al., 1996) use standard figures for mineralisation which are soil type and temperature dependant. Other more simplistic models make estimates based on total soil N and some use incubation tests. The most commonly used analytical test is to use a hot KCl extract (Whitehead, 1981) or a modification of that technique. Following cereal crops spatial variation in levels of potentially available N have been shown to be low but this is not the situation where they would be most valuable. Potentially the greatest use would be following the application of organic manures, the incorporation of cover crops or on soils with high organic matter contents. Organic N extracted with $CaCl_2$ may also give an estimate of potentially available N (Appel et al., 1995).

Conclusions

Before, during and after the potato crop mineral nitrogen within a field can be variable as a result of variations in soil type, residues from previous crops and inaccuracies in fertiliser N application. A robust soil sampling programme can give a good representation of the area sampled but the user should always be aware of the potential for variations within the area sampled be it field or experiment. Once the soil has been sampled and taken to the laboratory for analysis the correct technique must be chosen, that is the one appropriate for the information required for the predictive model being used or the experimental aims.

Recommendations

Activity	In field	In experiments
Predictive sampling pre crop	At least 10 points according to established pattern, no more than 10 ha sampled	At least 10 points according to established pattern within each block
Predictive sampling in crop	As above but bearing in mind potential variations in Mineral nitrogen due to application errors and ridges	As above but bearing in mind potential variations in Mineral nitrogen due to application errors and ridges, each plot to be sampled
Post harvest for residual Mineral nitrogen	As above following cultivations for the following crop	As above, sampling may be needed immediately post harvest and post mineralisation of incorporated residues.
Analysis		
Total N	Kjeldahl or micro Dumas	Kjeldahl or micro Dumas
Mineral nitrogen	KCl extract	KCl extract
Potential Mineral nitrogen	Hot KCl or incubation	Hot KCl or incubation

3.5   Using soil nitrogen status in practice - the need for Decision Support Systems

M.A. Shepherd

Chapter 3 described nitrogen cycling in the soil, the sources of nitrogen in soils and how best to sample fields for soil nitrogen measurement. Here, we answer three questions: what is meant by 'optimum nitrogen rate'? What role does soil N measurement play in fertiliser advice? Is there a need for decision support systems for fertiliser N?

Introduction

Attempts to 'optimise' nitrogen fertiliser requirements for potatoes (or for any crop) are hampered by large differences between sites and years in both mineralisable native organic nitrogen and inputs of fresh organic nitrogen (Ris et al., 1981). Many factors affect soil N supply, as described earlier (Chapter 3). Because mineral N levels can fluctuate between years and between sites, measurement is seen as an advantage when planning fertiliser applications, and this is now a prerequisite of many European advisory systems. For example, Neeteson (1989) demonstrated an average reduction of 10% in fertiliser inputs with no loss of yield when basing applications on Nmin measurement compared with a standard dressing. However, factors such as soil-type and previous organic manure applications influence fertiliser requirement also (Neeteson & Zwetsloot, 1989). Therefore, decision support systems are often used to take account of these many factors.

A nitrogen fertiliser recommendation should be large enough to guarantee a high crop yield with a high fertiliser N recovery in the crop, and low enough to avoid unnecessary expense on fertiliser not needed for yield and so liable to loss by leaching (Neeteson, 1989). The definition of 'optimum nitrogen' is based on economic considerations. It is defined as the point on the yield curve at which the cost of additional fertiliser is not justified by the additional yield that will be obtained with the extra N. Therefore, an important consideration in any fertiliser recommendation system must be the price of the fertiliser and the price of the produce: increasing the value of the product or decreasing the fertiliser price will increase the amount of fertiliser that can be applied economically. Conversely, decreasing potato prices or increasing fertiliser price will reduce fertiliser inputs.

Our aims here are therefore to describe (a) the typical N response of potatoes and the factors affecting it, (b) the benefits of soil N measurement and (c) the implications of N fertiliser applications on profitability and the environment.

The concept of 'optimum nitrogen'

Defining optimum nitrogen

(a) Crops taken to maturity - Figure 1 shows the shape of a typical nitrogen response curve for potato crops *taken to maturity.* Unfertilised yield is derived mainly from soil nitrogen supply (with contributions from N deposited from the atmosphere and applied in irrigation water). The response curve shape shows a large increase in yield with applied N at low rates, a smaller rate of response at intermediate rates and declining to a small or nil effect from additional nitrogen at higher rates. Whereas some crops, such as cereals or sugar beet, tend to show a yield decrease with large N applications, potatoes generally reach a plateau without a negative effect.

(b) Crops harvested before maturity - It is important to stress that the response curve shape in Figure 1 is generally true only of potato crops taken to maturity: at earlier harvest dates, yields from larger application rates can be less than from intermediate rates. This is because N prolongs the growing season, so that large applications of nitrogen will only be used effectively if the crop is allowed to continue to maturity. If the crop is harvested early, yield and the nitrogen optimum will be less. Furthermore, N applications above the crop's needs could delay transfer of dry matter to the tubers, thus *decreasing* yield - then the yield plateau shown in Figure 1 would show a downturn. Clearly, anticipated harvest date is an important consideration for fertiliser policy.

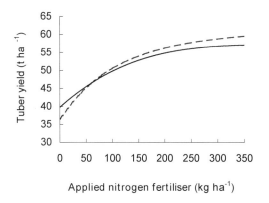

Applied nitrogen fertiliser (kg ha^{-1})

Figure 1   *Typical nitrogen response curve shape for potatoes from UK (unpublished, broken line) and NL (Neeteson, 1989; solid line) for crops taken to maturity.*

Some define the economic optimum as the nitrogen rate at which extra fertiliser will not produce sufficient additional yield to cover the cost of the fertiliser. This point is often close to the plateau of the response curve and so is difficult to determine accurately (Neeteson & Wadman, 1987). It could be argued also that, because the optimum depends on produce price, it will only ever be known after harvest and is therefore a redundant concept.

Crop factors affecting optimum nitrogen

Several factors affect the crop's requirement for nitrogen fertiliser and, so, the optimum nitrogen rate:

1. *Length of growing season* - This will generally determine the nitrogen uptake potential of the crop since early lifted crops will utilise less fertiliser than a crop taken to maturity. An important implication of this is also that large N rates can produce lower yields than intermediate N rates when crops are not taken to maturity because of fertiliser effects on setting tubers.

2. *Growth restrictions* - Efficiency of fertiliser use (i.e. the proportion of the applied N taken up by the crop) is a key influence on nitrogen fertiliser rate. This can differ between crops and is perhaps one of the factors most difficult

to assess. Clearly, pest and disease should be minimised but, also, unirrigated crops on drought-prone soils are more likely to use N less efficiently in dry years.

3. *Product/fertiliser prices* - Since Nopt is based on a standard ratio of product and fertiliser prices, changes in either will affect the recommended N amount. However, the shape of the response curve (Fig. 1) is such that fluctuations have to be large to warrant adjustments to nitrogen rates. Data from Neeteson (1989) can be used to show that increasing the price of fertiliser by 25% (or decreasing potato value by 25%) affected optimum N by < 8%. However, there may be situations where the market price is expected to be high (baking potatoes, for example) and then there might be justification in adjusting fertiliser rates.

Economic penalties

So, judging the crop's optimum nitrogen requirement at the start of the growing season is difficult. This is where a robust decision support system will help. As a guide, N rates should be such that final yield is close to the start of the yield plateau and certainly before there is any down-turn in yield which might occur after the yield plateau. This is because, from an economic viewpoint, under-fertilising causes a yield loss and over-fertilising wastes fertiliser (and can also cause a yield loss). Figure 2 uses published yield data to calculate the financial penalties from over- or under-fertilising. For many potato crops taken to maturity, over-fertilising has no negative effect on yield (Fig. 1) - unlike the case with cereals which might lodge, for example. However, If the potato crop is not taken to maturity excess N might also reduce yield (by delaying the dry matter partitioning to tubers): then the financial penalty will clearly be greater than that from wasted fertiliser alone.

Environmental penalties

Wasted fertiliser has important consequences for the environment. Figure 3 shows the general relationship between nitrogen fertiliser applied, yield and

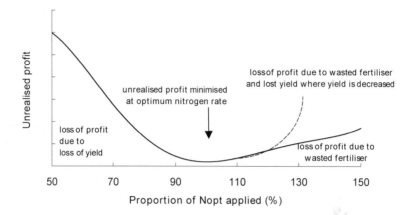

Figure 2   *Trend in unrealised profit arising from under or over-fertilising, based on data from Neeteson (Neeteson, 1989).*

post-harvest soil mineral N (i.e. nitrate in the soil in the autumn that is potentially available for leaching during the winter, depending on soil type and rainfall).

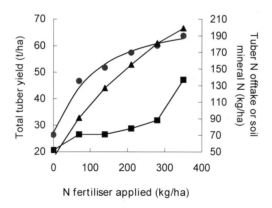

Figure 3   *Effect of N fertiliser rate on total tuber yield (circles), N offtake in the tubers (triangles) and post-harvest soil mineral N (squares): Shepherd, unpublished.*

Whereas fertilising up to Nopt has a relatively small effect on soil mineral N (and leaching risk), supra-optimal fertiliser applications have large effects. This risk therefore has to be avoided for environmental as well as economic reasons.

Using soil nitrogen measurements

Numerous experiments have shown that the larger the supply of soil mineral N at planting, the smaller the need for fertiliser nitrogen (Fig. 4). However, the relationship is not straightforward enough to be used as a simple decision support system without taking many of the factors, described earlier in this Chapter, into account. Chapter 5 provides more detail about individual recommendation systems.

There are two soil nitrogen fractions that are important for judging fertiliser requirement: soil mineral N and organic N (see Chapter 3.1). However, as well as the amounts of mineral N present, it is important to consider the distribution down the soil profile because deeper N will be used less efficiently by the potato crop. Mineral N at planting is affected by previous crop, soil type and winter rainfall: after a dry winter soils contain more mineral N because less has been leached. Similarly, less is leached from heavier textured soils. Soil mineral N is also influenced by manure applications and the quantity of soil organic matter.

Mineral nitrogen (Nmin)

Measurement of Nmin (predominantly NO_3^-) provides a 'snapshot' of the amount of plant available N in the soil at the time of sampling, providing all of the practical difficulties described in section 3.4 can be overcome. Many European advisory systems use this as a basis (described in Chapter 5). The main issues that need to be addressed are:
- time of sampling
- depth of sampling and weighting applied to each soil layer
- soil analytical method

We deal with these in later Chapters.

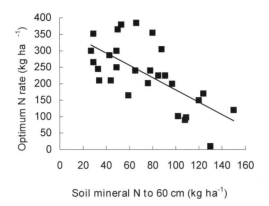

Figure 4   Typical relationship between pre-planting soil mineral N and optimum N for potatoes (adapted from Ris et al. 1981).

Organic nitrogen (Norg)

Nitrogen mineralised after the measurement of Nmin also needs to be accounted for. This contribution can be significant in soils treated with organic manures for example. In practice, it is one of the most difficult components of the recommendation system to assess correctly. Mineralisation, as described in Chapter 3.1, is a complex process affected by many factors.

We have many methods of estimating/measuring/predicting mineralisation of organic (Jarvis et al, 1996), including:
- measurement of soil organic matter or total nitrogen content
- anaerobic or aerobic incubations at fixed temperature and moisture regimes
- electro ultrafiltration (EUF)
- chemical extractants such as boiling KCl

Each has advantages and disadvantages but the greatest difficulty is relating any laboratory analysis to kg N ha[-1] mineralised during the growing season. This is discussed in more detail in Chapter 5.

The need for Decision Support Systems

We have shown that the nitrogen requirements of a potato crop depends on many factors. These can be separated into two groups: those which affect the total N uptake and rate of uptake of the potato crop and those which affect the soil nitrogen supply to that crop. Any shortfall has to be met by fertiliser. However, there can be large differences between fields in many of these factors. Soil nitrogen supply is one example, with differences caused by previous cropping and manuring. Then an 'average' fertiliser recommendation would not take account of these differences, thus failing to meet economic and environmental goals.

It is clear, therefore, that there is a need to provide field-specific fertiliser recommendations, and Decision Support Systems which take account of soil, crop and weather factors will help to achieve best fertiliser practice. Such support systems are described in later Chapters.

Conclusions

Many factors affect the nitrogen requirement of the potato crop, not least soil N supply. There is a need to take this into account when formulating fertiliser recommendations for a crop. The aim must be to generate field-specific recommendations to
- optimise profit
- minimise adverse environmental impact

Decision support systems will aid the farmer or adviser to achieve these objectives: fertiliser applications that do not take account of soil and crop factors are unlikely to consistently achieve these objectives.

Recommendations

1. Whereas fertilising up to Nopt has a relatively small effect on soil mineral N (and leaching risk), super optimal fertiliser applications have large effects. Decision support systems should be used for environmental as well as economic reasons.

2. Numerous experiments have shown that the larger the supply of soil mineral N at planting, the smaller the need for fertiliser nitrogen. Measurement of mineral N should therefore be considered as a starting point for fertiliser planning, especially where soil N supply is likely to be 'atypical'.

4 Plant and soil water status

4.1 Plant and soil water status: what is their role and what can we do with their values?

L. Dalla Costa & D.K.L. MacKerron

In the following chapter attention is focused on 'water', and the water relations of the potato plant will be discussed. A supply of water, stored in the soil, is potentially available for the plant to take up: How much water must be there, and how 'available' must it be for the plant to grow without stress? Water is absorbed by the roots and conveyed to the stem; but how far do they explore? Water then flows through the stem to the leaves to be lost to the air through the stomata: What, then, is the plant water status? Each of these steps in the flow of water can be studied and measured: But which parameters should be chosen best to describe the plant water status? How good are these values in describing the water needs of the plant?

Introduction

Water is key in determining potato plant development, growth and yield and the crop is widely recognised to be sensitive to water stress. There is an extensive literature describing the effect of even moderate water shortage on yield and quality, showing that the severity of effects depend on the time when water stress occurs, its duration and intensity. The results in many papers differ only in detail according to cultivar, soil characteristics, and weather.

The description of the water status of a plant requires us to think of a transpiring potato plant as part of a complex system involving

1) the soil and the water it contains,
2) the plant with its roots and leaves connected by the vascular system that conveys water from the roots to the transpiring leaf, and
3) the atmosphere, into which the water evaporates (is transpired).

The water in the three compartments has two principal characteristics, its energy state and its quantity. Quantity is straightforward. It describes how much water is there. The energy state of the water, called its water potential, describes its tendency to move from one place to another. It always moves from a place of higher energy state (higher water potential) to a place of lower energy state. – Water moves downhill! These two concepts lead on to a third one, this time characterising the conducting system. This is called the resistance to flow and it describes the difficulty with which the water actually moves. So, the soil offers a resistance to water flow towards the roots; the stem offers a resistance to flow towards the leaves.

These three apparently separate compartments are actually part of an integrated system, called the soil-plant-atmosphere-continuum (SPAC) in which there is a continuous flow of water between the compartments.

An idealised model (Figure 1) is used to visualise the water relations in a potato plant, and to identify some parameters that can be measured (in Chapter 4.2) to describe water status in the plant and in the soil, and to assess availability of and requirements for water. In Figure 1, three compartments are represented: the soil with its water potential, Ψs, the plant with its associated water potentials (indicated by root Ψr, stem Ψx and leaf Ψl), and the atmosphere with Ψa. Within and between these compartments there are resistances. Water encounters resistance in moving from one layer of soil to another and from the soil to the root. There are resistances to flow

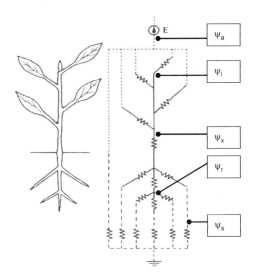

Figure 1   *Schematic model of the water relations between plant and soil, the water potentials, Ψ, and the resistances to flow.*

between root and stem and between stem and leaf. And, finally, there is a resistance to water movement from the leaf to the air. That last resistance is partially controlled by the stomata and is referred to as stomatal resistance.

Water flows from sites of 'high' water potential, (close to zero) in the soil, to sites of low water potential (more negative) in the plant and then to the atmosphere with even lower water potential. The driving force for this flow is the difference in water potential between compartments. That is to say that the plant draws harder on the soil water when the plant is water-stressed than it does when it is well supplied with water.

Soil water status is described both by how much water is stored in the soil, its water content, and on the energy required to draw this water from the soil (its water potential). Soil texture and structure mostly determine the relation between the water content of the soil and its water potential.

The water status of the plant results from the balance between the supply of water from the soil and the demand for water by the atmosphere. The amount of water available to the plant at any time depends on how much water is still stored in the soil, how strongly it is held by the soil particles, and on how deep and branched the root system is to abstract the water. Although there are resistances to flow within the plant, these are much smaller than the resistances to flow in and out of the plant so, if water is being absorbed by the roots, the water status of the leaves depends principally on their ability to hold the water against the atmospheric demand, through stomatal control.

Each of the two ways to describe the water status of the soil and the plant has its own merit. The quantity of water, grams of water per gram of dry soil or gram of plant dry matter can be relatively easy to measure. It tells how much water is in that part of the system and, more importantly, it indicates how much water is needed to restore the system to being well supplied. However, it tells very little about the strength with which the water is held, and energy required to abstract the water from that compartment. That is described by water potential.

Whenever there is an imbalance between the supply of water, provided by the soil, and the demand for water, determined ultimately by the weather, the plant may become water-stressed. Water-stress is to be avoided as it leads to reductions in yield and quality. The first effect of water-stress is a reduction in the rate of leaf expansion. At greater water-stress, stomata close in an attempt to conserve water but with consequent inhibition in CO_2 fixation and assimilate production. If the water-stress is prolonged and becomes more severe it can lead to wilting of the leaf and, ultimately, shoot death.

Water potential: a way to describe water status

Water potential, indicated by Ψ, is defined as the free energy in the system relative to the free energy (taken as zero) of pure water at the same temperature and at pressure of 0.1 MPa. It can be expressed in J mol^{-1} or J kg^{-1}, but is usually expressed in units of pressure, as KPa and MPa.

The total water potential of a compartment, Ψ without any suffix, comprises several component parts. In plants and soils, there is a component due to the materials that are held in solution, the solute potential Ψs. There can also be a gravitational potential, Ψg, for water held at considerable depth in soil or in tall plants. In soils alone, there is a component due to capillary and other absorptive forces between the soil particles. That is called the matric potential, Ψm. In plants, alone, there is a fourth component, the pressure or hydrostatic potential, Ψp, that develops when water is drawn into plant cells in response to the solute potential and causes them to swell to fill the space inside the cell walls. It is an effect that is also known as turgor and tends to balance the effect of the solute potential. All the component potentials described have negative values except the pressure potential, which is normally positive.

The balance of all the component water potentials can be expressed in a simple equation:

$$\Psi = \Psi m + \Psi s + \Psi g + \Psi p \qquad \text{Eqn. 1}$$

Water in the soil

Every soil has a unique set of curves that relate the soil water content (amount) to the soil water potential. One outer curve describes the relation between content and potential as the soil dries from a fully wetted condition and another describes the same relation as the soil is wetted from dryness. Between the envelope produced by these two curves is a whole family of subsidiary ones describing the relations as the soil dries from partially wet, or wets from a partially dry condition. These are known as the soil water characteristic curves, or soil water retention curves. They differ markedly between soils with differing soil textures. (Figure 2). At any given water potential, different amounts of water are held by soils with differing texture. (Table 1).

Field capacity and permanent wilting point are two loosely defined but usefully characteristic levels of soil water content. The first, *in situ* field capacity refers to

the amount of water present in a thoroughly wetted soil after downward
drainage has occurred and the water content has become relatively stable. This
characteristic represents the maximum quantity of water that the soil can retain
and store. The equilibrium after downward drainage is reached quickly in sandy
soil and quite slowly in clay soils. For this reason, the definition of field capacity
is often given by reference to the soil moisture release curve. It is commonly
defined as the water content at 10 KPa.

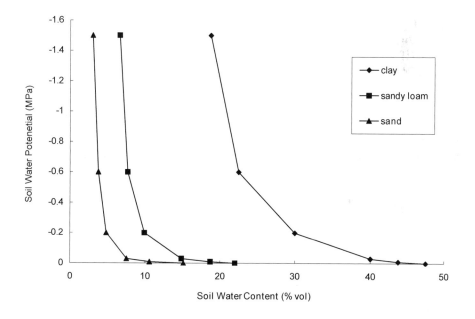

Figure 2   *Relation between soil water potential and volumetric soil water content for three*
              soil types.

Table 1   *Water available between field capacity and two levels of soil water potential in soils*
of differing textures, expressed as % of soil volume (Burton, 1989). Where irrigation
is scheduled using soil water potential, -0.04 MPa might be taken as the trigger
point for irrigation. Water potential in unirrigated crops will generally fall below –
0.1 MPa, by which stage plants are significantly stressed and growth is impaired.

Type of soil	Threshold soil water potential	
	-0.04 MPa	-0.1 MPa
Fine sand	4	6
Sandy loam	6	10
Silt loam	6	11
Silty clay	7	13

At the other end of the scale, as the soil dries out due to water uptake by the crop, the plants reach a stage where they can no longer obtain as much water from the soil as they are losing to the atmosphere, and they wilt. Initially they will recover turgor overnight while the evaporative demand of the atmosphere is very low. But shortly after that, unless the soil is watered, the plants will not recover their turgor. Then the soil is said to have reached the permanent wilting point. Again, the definition is loose, being influenced by soil type, tillage, and atmospheric humidity. So, as for field capacity, the permanent wilting point is often given by reference to the soil moisture release curve. It is commonly defined as the water content at 1.5 MPa. Between these two limits of soil water status, potatoes and other plants will be more or less stressed depending upon the weather conditions, and have a range of responses. Effects of water-stress on leaf expansion have been reported starting at -25 KPa or even wetter.

The amount of water between these two extremes, field capacity and permanent wilting point, is known as the available water content. So, in the layer explored by the roots, there is a certain amount of water available to the plants, the plant available water, determined by root depth and the moisture release characteristics of the soil.

Soil water is readily available if the soil is close to field capacity, but it is increasingly difficult to extract when the moisture approaches the wilting point. Some authors have proposed that 50% of the plant available water is readily available. Others have proposed a value of 75%. When the crop has to rely on the water that is not readily available, there are progressively severe restrictions of transpiration, photosynthesis and growth.

The amount of water that is readily available may be estimated by the equations:

$$Va = 0.5 \times (Vfc - Vpw) \times Dr \qquad \text{Eqn. 2}$$

Or

$$Ma = 0.5 \times (Mfc - Mpw) \times \rho \times Dr / 100 \qquad \text{Eqn. 3}$$

Where:
Va	=	Available water content (in mm)	
Ma	=	Mass available water content (in kg m^{-2} which is numerically equivalent to mm)	
Vfc	=	Volumetric water content at field capacity	
Mfc	=	Mass water content at field capacity (%)	
Vpw	=	Volumetric water content at wilting point	
Mwp	=	Mass water content at wilting point (%)	
Dr	=	rooting depth (in mm)	
ρ	=	bulk density of the soil (in g cm-2)	

Thus, in a soil with 25.5% and 10.5% by mass of soil water at field capacity and wilting point respectively, 1.3 g cm^{-2} bulk density, and 700 mm estimated root depth, the water initially available to a plant would be 68.3 mm.

Water status in the plant

The water potential of the plant is also described by equation 1. Generally, the plant is at a lower water potential than the soil and draws water from it. The value of the plant water potential is critically dependent on soil water potential and on the values of the resistances to water flow in to the roots, and out through the stomata. It fluctuates diurnally, being lowest in the afternoon. At night, when evaporative demand is almost zero, the plant will rehydrate, and its water potential rises to approach that of the bulk soil.

Plant water potential is not easy to measure, and so generally other measurements are made that reflect other aspects of the plant's water status. These include stomatal resistance (by porometry), and difference in temperature between the crop canopy and the air caused by transpiration (by infrared thermometry) and relative water content. These measurements are considered more fully in Chapter 4.2.

Within the plant there is little difference in water potential between roots, stems and leaves because resistances to flow are small. Significantly, the tuber sits like a buffer adjusting to changes in the rest of the plant. Occasionally, when transpiration is high, or changing rapidly, there can be large disparities between leaf and tuber water potential.

The levels of the several components of water potential inside the plant are crucial for several physiological processes: leaf enlargement, stomata opening, tuber enlargement, and photosynthesis. However, the rate of leaf appearance is unaffected by water-stress (Jefferies, 1989). Leaf growth rates are highly correlated with afternoon measurements of leaf water potential, Ψ , and pressure potential, Ψp, and have been reported to be reduced in comparison with that in irrigated crops when Ψ is less than -0.28 MPa and Ψp is more than $+0.5$ MPa (Jefferies, 1989). Leaf extension stopped completely when Ψ was -1.13 MPa or less and the pressure potential, Ψp, of leaves was close to zero where Ψ was -0.8 MPa or less. Vos & Oyarzun (Vos & Oyarzun, 1987) reported a sharp increase in stomatal resistance following a small decline in Ψ from -0.5 to -0.6 MPa, and lessening response at lower Ψ as the stomata became progressively closed. This led to a fairly direct relation between Ψ and photosynthetic rate over the measured range of -0.5 to -1.05 MPa. Other experiments (Jefferies, 1993) showed that reductions in final size of leaves by water-stress were due to lower rates of expansion rather than to shorter periods of expansion. Each leaf expands during a defined period and so, even relatively short periods of water stress can have permanent effects on the area of leaf that is formed. So, there are other reports (Gandar & Tanner, 1976) of total leaf area being reduced by as much as 60%, without later compensatory growth, where one week's irrigation was omitted. The same study (Gandar & Tanner, 1976) found that leaf and tuber growth ceased at similar water potentials, around -0.5 MPa. At lower water potentials, the tubers shrank until the water-stress was relieved. Translocation of assimilates to the tubers and tuber growth occur out of phase with photosynthesis, taking place mostly at night, when the plant is rehydrated and turgor is positive. Both the uptake of sucrose into tubers and its conversion to starch are sensitive to osmotic potential (Oparka & Wright, 1988).

The rehydration of a water-stressed plant overnight is important for the continued functioning of the plant. In the field, the rate of rehydration is limited by the rate of movement of water within the soil. Obviously, as the plant abstracts water during the day, it makes the soil around its roots drier. However, the hydraulic conductivity of soil decreases rapidly as it dries. So, the rate at which the rhizosphere re-wets from the surrounding bulk soil is the limiting factor in the plant's water economy. The presence of deep roots may improve the survival of a plant by allowing access to more water but the additional amounts of water that are accessed at lower depths has a limited benefit. In a simulation study (Jefferies, 1993), Jefferies found that increasing the rooting depth by 20% gave a 7.3% increase in yield at 175 days after planting under a severe drought cycle. It had no benefit in shorter growing seasons or less severe drought.

Leaf relative water content (RWC), which is the ratio of the amount of water in a leaf to the amount that it would hold when turgid, is a measure of the hydration of a leaf. It is related to leaf, solute and pressure water potentials (Ψ, Ψs, and Ψp) but we will not describe the relations here. There are arguments that RWC is the best measure of plant water-stress but opinions are divided on this. The measurement of RWC will be explained in Chapter 4.2. Vos & Groenwold (1988) gave several characteristics for the relations between Ψ, Ψs, and Ψp in potato, summarised here in Table 2.

Table 2   *Characteristics of water relations pooled from well-watered and droughted plants,*
of three varieties, at three dates of measurement (Vos & Groenwold, 1988).

Parameter, conditions, and units	Mean value
Full turgor (RWC = 1)	
Ψp (MPa)	0.67
Slope of Ψp x RWC (MPa)	8.4
Wilting point (Ψp = 0)	
Ψ (MPa)	-0.86
RWC (fraction)	0.92

The root system

The potato plant is generally said to be shallow rooting but this is to ignore the many published reports of potatoes rooting to 80, 100, and even 150 cm; (for examples, Vos & Groenwold, 1986; MacKerron & Peng, 1989). The attributed shallow rooting may indicate greater sensitivity to adverse soil conditions such as roots confined to the plough layer (Gregory & Simmonds, 1991) and to the preponderance of fine roots. As the majority of potato roots have been found to have diameters less than 0.2 – 0.24 mm (Vos & Groenwold, 1986). - The depth of the root system and the rate at which it explores the soil are important factors determining the amount of soil water available to the plant. A number of published reports on rooting in potato have been collated (Gregory & Simmonds, 1991) and are summarised at Table 3.

The distribution of the roots in the soil profile is also important. Although there is a report (Lesczynski & Tanner, 1976) of the maximum amount of root being at 15 – 20 cm, that soil appears to have had a plough pan. Other studies (Vos & Groenwold, 1986) found maximum rooting at the bottom of the ridge, at 20 cm. In contrast, maximum root densities have been found at between 60 and 80 cm in both wet and dry conditions, where a compost was used designed

Table 3   *Rooting depth observed in experiments carried out on several cultivars on a range of soils (Gregory & Simmonds, 1991).*

Soil	Cultivar	Rooting depth (m)
Marine clay	Bintje	0.8-1.0
Sandy clay loam	Majestic	0.90
Sandy clay loam	Désirée	1.0
Sandy loam	Record	0.80
Sandy loam	K. Edward	0.47
Sandy loam	Majestic	0.55
Sandy loam	M. Piper	0.50
Sandy loam	P. Crown	0.60
Loamy sand	Russet Burbank	0.70
Loamy sand	Vanessa	0.90

to allow the plants to fully express their potential for rooting (MacKerron & Peng, 1989). That study also found that there was far more root where the soil was well-watered.

The rapid early growth of roots is followed by a period in which a proportion of the roots die. That can occur early, beginning at 50 – 60 days after emergence or not until the onset of senescence; for example, at 89 days after emergence in frequently irrigated potatoes (Lesczynski & Tanner, 1976). In the latter study, root death had reduced the size of the root system by 30 – 50% by 107 days after emergence. However, there is evidence that root growth continues after the start of root death, so that large roots systems extending to 150 cm have been found in September, 100 days after emergence (MacKerron & Peng, 1989).

Although the partitioning of dry matter between shoot and tuber is affected by drought treatment, root dry matter seems to be a conservative fraction of the whole (MacKerron & Peng, 1989). So, just as the total dry matter production is reduced by water-stress so, too, is the production of roots.

Water in the atmosphere

Water vapour in the air exerts a pressure, known as vapour pressure. If a sample of air is sealed over a surface of pure water under conditions of constant temperature, it will eventually reach a state of equilibrium, when the amount of molecules evaporating from the water surface is equal to the molecules entering the liquid. This stage is named saturation vapour pressure, and it is temperature dependent. Actual water vapour pressure present in the atmosphere is generally lower, and it is never saturated. The difference between the ambient vapour pressure in the atmosphere and the saturated vapour pressure is termed vapour pressure deficit, or saturation deficit. Vapour pressure ranges between -1.4 MPa when relative humidity (RH) is 99% (almost saturated) and for example -310MPa when RH is 10% (very dry).

Water quality

The increasing use of water for human needs, the competition among different agronomic uses, including demand for other crops, limits the availability of water and lowers its quality. The quality of water has different aspects: temperature, pH, particles suspended, particular ions (salinity), and other pollutants. The lowered quality of water available for agricultural uses is mainly due to its high levels of heavy metals and other toxic compounds but in the Mediterranean countries, salinity is regarded as the main problem. (See for example Levy, 1992).

Conclusions

The functioning of the potato plant is very sensitive to the supply of water. That sensitivity can be observed in the several effects of any disruption in the movement of water through the continuum of the soil- plant-air. This sub-chapter has described the processes and states that are most sensitive and has shown they ways in which they change under water-stress. Methods for measurement of these changes are considered in the next sub-chapter. Part of the vulnerability of of the potato to water stress is a reflection of its root system. That system is frequently said to be shallow but is better described as being sensitive to poor soil conditions and tillage. The ridge of soil in which the potato is grown can be the most immediate source of both water and nutrients, but it is important for the well-being of the crop that it has access to greater depths. The extension growth of the leaves is the process that is most sensitive to water-stress, such that there is almost no upper level of soil moisture that can guarantee potential growth. Further, the stomata of potato leaves begin to close at relatively high leaf water potential (low levels of stress), reducing transpiration and photosynthesis. Consequently, growth and yield will be seriously reduced whenever 50% or more of the available moisture in the soil has been taken up.

Recommendations

Important soil characteristics like field capacity, available water content, and wilting point should be known for each soil on which potatoes are grown. Then, soil water status should be followed carefully, and action taken to recharge soil water by irrigation if water-stress is to be avoided.

Due to the inability of the rooting system to penetrate a plough pan or an indurated layer, tilling should done carefully to break up any impeded soil layers, and to avoid compaction of the soil. For optimal plant growth the water available in the soil should never fall below 50%.

For optimal leaf growth, expansion and active photosynthesis leaf water potential should never drop below -0.4, -0.5 MPa. However, it must be recognised that diurnal fluctuationi in water potential can cause the plant to exceed that threshold temporarily during the hours of greatest evapotranspiration.

Chapter 4

4.2   Measurement techniques for soil and plant water status

S. De Neve, D. MacKerron, & J. Igras

In this chapter we look in some detail at the problem of measuring the water status of soil and plant, soil water balance, and the meteorological factors driving the crop demand for water, evapotranspiration. Questions that will be answered include: How can we measure the crop water status? How can the soil water content and soil moisture tension be measured, and which method is better? Which methods are available to estimate evapotranspiration? Which approach is better - to measure crop water, soil water or estimate evaporation? Which measurements are necessary? Which measurements are the best to make?

Introduction

A crop needs water to grow, and anything less than a proper supply of water can reduce yields dramatically. This is why many crops are irrigated and why it is important to be able to monitor soil or crop water status. Soil water content changes continuously, as water is withdrawn by the crop and replenished by rain or irrigation. The crop water status changes to reflect the balance between the supply of water (dependent on weather and physical soil properties) and the evaporative demand (dependent on weather and crop conditions). A good measure of the crop water status will indicate the degree of water stress the crop is experiencing, whereas a good measure of the soil water status gives indirect information about how well the crop may be provided with water.

In Chapter 4.1 the consequences of water stress on crop growth were discussed and some definitions were given of key parameters of soil and crop water status and their critical values. In Chapter 4.3 attention is given to sampling strategies for measuring plant and soil water status considering both spatial and temporal variation. In this chapter we give an overview of techniques that can be used to measure plant water status, soil water status, and evapotranspiration and we indicate the merits and disadvantages of each.

Methods

There are several approaches that can be taken towards assessing the adequacy of water supply to the crop. First, aspects of the crop water status can be measured. This has the advantage of giving direct information about the crop. Its principal disadvantage is that it is unlikely to indicate impending water-stress in the future. By the time that the crop has become measurably water-stressed, some potential yield has probably been lost. Nonetheless, we include descriptions of methods to measure crop water status as they can provide useful additional information, particularly to advisers and researchers seeking explanations of effects. A second approach is to measure aspects of the soil water status. The disadvantage of the approach is that it gives only indirect information on the crop. This is generally outweighed by its advantages: the measurements are usually less variable, and are easier to make and interpret than measurements of the crop water status. In addition, measurements of soil moisture status can give advance warning of impending limitations to the water supply. The third approach is to measure neither plant nor soil water status but to use a combination of measurement of inputs (rain and irrigation) with calculation of losses (evaporation and drainage) to maintain a soil water balance. It is however highly recommended to make additional measurements of soil moisture to increase irrigation efficiency and reduce risks of yield losses. We will describe this approach also.

Assessing crop water status

As discussed in Chapter 4.1, measurements are available for the (absolute) amount of water in the plant , its water potential (how strongly the plant is drawing on the soil water), and other indirect estimates of water status. We discuss the more important ones and indicate the fields of work for which they are most suitable.

Water content and water potential

The 2 above mentioned methods allow direct determination of the water status of the plant. The relative water content (RWC) gives the (absolute) amount of water in the plant relative to what it holds at full turgor. Water Potential can be measured directly by pressure chamber or indirectly by use of psychrometers (SoilMoisture Equipment Corp., California, USA). Although both techniques are simple, especially the RWC determination, they are mainly research tools and can not be used directly to guide irrigation scheduling.

Water use

Stem flow

Stem flow gauges (e.g. Dynagage, Flow32™) are sensors that allow direct measurement of the water used by plants as it moves up the stem. The potato stem has a strongly angled cross-section that gives difficulty in effecting and maintaining a good thermal contact between stem and sensor. The sensors are sensitive but expensive. They also require sophisticated data logging facilities to record and process the measurements. Although the suppliers (Dynamax Inc., USA; ADC Bioscientific Ltd., UK; Delta-T Devices, UK) offer software enabling them to be used in crop water management, the expense of the system is a great disadvantage for commercial use.

Porometry

Stomatal aperture is the dominant factor in the diffusive conductance of leaf surfaces, which controls both the water loss from plant leaves and the uptake of CO_2 for photosynthesis. Measurements of diffusion conductance are therefore important indicators of plant water status and provide a valuable insight into plant growth and plant adaptation to environmental variables. A porometer is an instrument that measures the stomatal conductance or resistance. There are two general designs, equilibrium and dynamic, that differ in their control philosophy. In both designs, a leaf or leaflet is either clamped inside a chamber or the chamber is clamped to the leaf. Water loss from the enclosed leaf then tends to increase the humidity in the chamber. In the equilibrium porometer, dry air is added at a measured rate to maintain a stable humidity close to the ambient conditions. In the dynamic porometer, the chamber is purged with dry air and the transit time is measured while the leaf wets the air between two threshold values, dry and wet, either side of ambient conditions. The instrument cycles repeatedly through drying and wetting phases.

The dynamic porometer is the most common form of porometer, having been considerably miniaturised and automated. The calibration, too, is automated and the latest model (Porometer type AP4, Delta-T Devices Ltd.) gives a direct reading of stomatal conductance or resistance.

The porometer is an excellent research tool enabling both measurement and interpretation of plant water status but it is not suited to the routine management of crop water supply.

Infrared thermometry

Dry surfaces exposed to sunshine are warmer than the air around them but wet surfaces can be cooled by evaporation. Where stomata are open, water evaporates from the leaf and so cools it. The greater the evaporation rate, the greater the cooling of the leaf. Where the supply of soil water is limited, less water is evaporated and there is less cooling of the leaves and the canopy temperature is higher than one would expect for a crop with an adequate water supply. That is, the leaf-air temperature difference is less than it would be in an unstressed crop.

Using the humidity of the air, and the leaf-air temperature difference, one can calculate a Crop Water Stress Index (CWSI) that can be used to schedule irrigation (Ben-Asher et al., 1989). Critical values of the CWSI (i.e. values where yield reduction can be expected) can be derived from field measurements on crops growing under a range of water stress. This technique is now reaching beyond development and is approaching the stage where it can be used in commercial applications.

To use the IRT method (Infrared Thermometry) for calculating daily crop evapotranspiration, the following equipment is required: A portable commercial quality infrared thermometer capable of measuring crop canopy surface temperature, a PC or compatible microcomputer, access to weather data, a current copy of IRT Et software (Centre for Irrigation Technology).

Assessing soil water status

More of the techniques for measurement of soil water are suitable for commercial application and so, assessing soil water status is a more common approach to managing the water supply to crops. As for crop water status, measurements are available for how strongly the soil retains water, its water potential, and for the amount of water in the soil. Optimizing water supply will not only optimize water supply to the crop, but will also increase fertiliser efficiency by enhancing fertilizer uptake and reducing losses by leaching or denitrification. There are both direct and indirect measurements and here we consider the more important ones and indicate the fields of work for which they are most suitable.

Water content

Gravimetric method

This is a direct method and is the standard against which all other methods are calibrated. Fresh soil samples are weighed, placed in a drying oven between 100 and 110°C and dried to constant weight. The moisture content is calculated from the water loss and may be expressed on a mass basis or, if the samples were taken carefully with known volume, the water content can be expressed on a volume basis. Drying time is usually between 10 and 24 hours, depending on the type of oven. Although this method is very simple, some problems can arise. At drying temperatures above 100°C water can be lost from clay colloids and there may be some loss of organic matter. The two main problems with the method are that it is labour intensive and so, expensive, and that it is destructive. Repeated measurement requires other samples to be taken. Spatial variability demands that a large number of samples are taken on any occasion.

Water in the soil, modifies many of its physical properties. This gives scope for indirect measurement of soil water. All indirect methods require calibration. Depending on the property being measured and the accuracy required it can be possible to accept manufacturers' calibrations, or it may be necessary to do one's own. Once the indirect methods are calibrated for a given soil and probes are installed, continuous measurements of soil water status can be made with minimum additional effort and costs.

Neutron probe

This method is based on the moderation or slowing down in soil of fast or high energy neutrons emitted by a radioactive source and counting the density of 'slow' or low energy neutrons. In soil, the principal moderator is the hydrogen atom, and changes from time to time are caused by changes in the water content. The calibration is dependent upon soil type. Strictly, count rates should be calibrated against volumetric water content determined gravimetrically on each soil to be measured. However, the interesting measurement for managing the water supply to a crop is the difference between readings taken at successive times. The errors from using a 'standard' calibration are less for changes in soil water than for the absolute value. Therefore, in commercial practice, the lower accuracy from using standard calibrations may be acceptable.

Using a neutron probe involves the prior installation of aluminium access tubes to allow repeated measurements at each location. This needs to be carried out very carefully to obtain reliable measurements.

Care has to be taken when measuring in the upper soil layers because of the escape of fast neutrons. This is both a hazard and a cause for separate calibrations in the upper layers.

Depending on the soil moisture content, the measurements are influenced by a sphere of soil of radius 7- to 35-cm . It is not possible, therefore, to detect a sharp change in soil moisture, as at a wetting front.

This is a field method that is widely used but has both advantages and disadvantages. The principal advantages are that it samples a large volume of soil – typically each reading represents a sphere of soil, 15 cm radius, which will reduce variability of the measurements as compared to point observations– the soil is measured in depth, and repeated observations are made on the same soil every time. The principal disadvantages of the method are the hazard of working with a radioactive source and, therefore, the associated precautions, and the expense of the equipment. There are crop consultants who will provide measurement of soil moisture by neutron probe as a service, at a cost. This allows the grower to avoid the two main disadvantages.

Time domain reflectometry (TDR)

Time domain reflectometry is based on the change in dielectric constant of the soil with changes in soil moisture content. Measurements, which take about 15 seconds, can be made manually, or the recording instrument can be programmed to make and store readings automatically. The equipment includes a cable tester, probes – wave guides - and a computer to record the wave forms, analyse them, and save the resulting calculated soil moisture. Optional equipment includes a multiplexer for automated multiple readings.

Calibration of TDR for measuring soil water content can be done easily in sample soils with a range of water contents. However, (Topp et al., 1980) developed a general calibration based on a range of soils (Fig. 1). For general purposes, this general calibration relation can be used but for research purposes, where more accurate moisture contents are required ($\pm$ 0.01 m^3 m^{-3} and better), calibrations should be done for every soil used.

The probes for the TDR system act as waveguides and are conveniently made from rods of stainless steel. Several different waveguide systems are available to be portable or for use permanently installed. The soil volume sensed by the probes is the full length of the conductors but only a small cross-sectional area. Therefore, measurements can be made with very high spatial resolution and very close to the soil surface. Where probes are buried

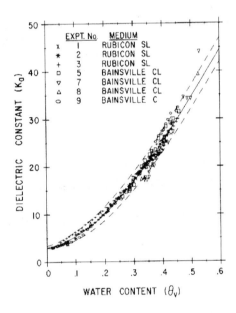

EXPT. No.		MEDIUM
x	1	RUBICON SL
*	2	RUBICON SL
+	3	RUBICON SL
□	5	BAINSVILLE CL
▽	7	BAINSVILLE CL
△	8	BAINSVILLE CL
○	9	BAINSVILLE C

Figure 1   The Topp et al. (1980) calibration relation between soil moisture content θ_v and the soil dielectric constant for a range of soils.

horizontally in the field measurements can be made at very small depth increments. Where, as is more normal, the probes are installed vertically, the reading corresponds to the full depth of the soil probed. It is only possible to estimate profiles, then, by using several sets of probes with differing lengths, e.g. 15- , 25-, 35-cm.

In recent years, TDR has been used extensively in research but it is also available for commercial use by farmers. Another advantage of TDR is the possibility for measuring simultaneously the soil bulk electrical conductivity. Monitoring the electrical conductivity of the soil can be useful in determining the leaching requirement when irrigating, and to warn the farmer for pending salt stress to the crop. Available systems include Tektronix, Trase System by SoilMoisture Equipment Corp., California, USA; Vadose by Dynamax Inc. and ADC Bioscientific Ltd.

For surface measurements probes can simply be inserted vertically, or buried horizontally in the topsoil and readings can be made. Where measurements are needed deeper in the profile, either, longer probes can be inserted using a special tool and taking care that the rods are parallel, or a pit has to be dug to allow the probes to be inserted horizontally into the soil at each depth, again taking care that the rods are parallel. After probe installation the pit is refilled with the original soil. When probes are installed as described here they should be long enough to minimize the influence of soil disturbance on the later readings.

Frequency domain reflectometry (FDR)

There are now a growing number of other instruments available on the market based on sensing capacitance and the dielectric constant of the soil. They tend to be stable in operation, lend themselves to automatic logging and even control. Capacitance probes may be installed directly in contact with the soil or probes are inserted in PVC access tubes.

There is now a wide range of types on offer for both research and commercial applications. These include:

The Theta probe (Delta-T Devices) (see Fig. 2) that measures volumetric water content of soil and other growth media to within 2% (to within ±0.02 over the range $0 - 0.5$ m^3 m^{-3}). The probes are easily pushed in to the soil surface or they can be buried. They are easy to calibrate for specific soil types, are frost resistant, and can be used in saline conditions. They need an excitation voltage and the output is accepted by most types of millivoltmeter or data logger.

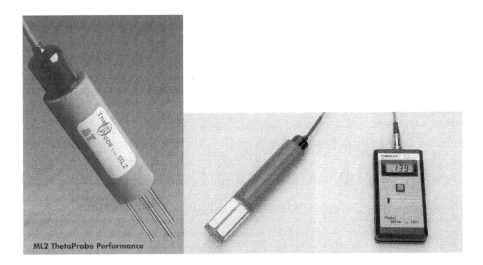

Figure 2   *The ML2ThetaProbe for measuring soil moisture content (left) and the equitensio-*
            meter for measuring soil water potential (right).

The Enviroscan system has probes with multiple sensors located at a range of depths that are adjustable. (Sentek Pty Ltd., Adelaide, South Australia. UK agent Peter White management). It has been shown to measure volumetric soil water contents accurately during laboratory calibration as well as in field conditions (Paltineanu & Starr, 1997; Starr & Paltineanu, 1998). A similar instrument called the Diviner is produced by Sentek Pty Ltd., Kent Town South Australia. It has a 10 cm radius of influence in the soil.
One version of such sensors has been developed to allow spatial averaging. The Aquaflex (Streat Instruments Ltd., Christchurch, New Zealand) incorporates the sensor in a robust tape, 3-metres long that is buried in the soil. With a 5-cm diameter of influence, it samples a volume of 5.8 litres.

Hillhorst & Dirksen (1994) listed some advantages and disadvantages of both TDR and FDR techniques. They concluded that TDR is more suited to research purposes, with high spatial resolution as a very important feature. Whereas FDR has superior features from an operational point of view: it is low cost, robust, reliable and easy to use. Finally FDR can be connected to a data logger at a distance of up to 500 m without problems (Paltineanu & Starr, 1997).

Water potential

The distinctions between water potential and water content were presented in Chapter 4.1. Here we are concerned only with how water potential can be measured and its value interpreted.

Gypsum block sensors are inexpensive but have low accuracy and stability. They are suited only to measure trends and have a short useful life. Don't use them.

Psychrometers

Soil psychrometers, measure the vapour pressure of the soil using thermocouples set in small ceramic cups. They can be very accurate in relatively dry soil, but are not reliable within the normal range of soil moisture in an irrigated system. This, too, is a research tool (SoilMoisture Equipment Corp., California, USA).

Tensiometers

Tensiometers indicate how much energy the plant must expend to extract water from the soil. They operate over a very limited range of water potential that is not nearly as wide as that tolerated by growing plants but is wider than the range tolerated by growers who wish to avoid restrictions to growth. That is, the range is adequate for monitoring the need for irrigation.

Modern tensiometers consist of an acrylic or metal tube full of water, with a porous ceramic bulb on the lower end that has to be kept in good contact with the soil. At the top end there is a means to refill the tube and, also, either a Bourdon pressure gauge or a piezo-resistive pressure transducer. The water filled tensiometers are well suited to irrigation monitoring and leachate studies in wetter soils. Their accuracy is good (±0.2 kPa over the range 0 – 85 KPa) but they need periodic refilling, degassing, and recalibration. The Theta probe (see capacitance sensors earlier, fig. 2) can be supplied with a calibration to allow the measurement of soil water potentials as low as -1500kPa with ±5% accuracy. In the 0 - 100 kPa range, accuracy is ±10 kPa.

Tensiometers are available in various lengths for monitoring water potential at a range of depths in the field (fig. 3). Those with a pressure gauge can be read directly. Those with a pressure transducer are read by connecting the transducer to a small digital display (millivoltmeter) or to a data logger. The electrical output makes this form suitable for continuous logging, 'remote' recording, and for automatic switching for irrigation control. Tensiometers are used in commercial horticultural applications for scheduling irrigation.

Instruments and suppliers include: Skye Instruments Ltd., UK; SoilMoisture Equipment Corp., California, USA; Delta-T Devices, UK.

Problems with tensiometers include: When dissolved gasses come out of solution and break the water column, the tensiometer gives a faulty, low reading. At higher elevations the operating range of the tensiometer will be reduced due to the reduction in atmospheric pressure. In common with several of the devices described here, tensiometers only give spot readings.

To install a tensiometer a hole should be augered of the same diameter, or a slightly larger than the barrel of the tensiometer. A slurry of soil is placed at the bottom of the access hole and the tensiometer is pushed down the hole and seated with a twisting motion. If the diameter of the access hole was larger than the diameter of the tensiometer then the space should be back-filled again with fine soil and compacted with a rod to prevent preferential

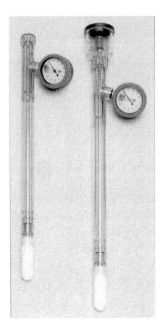 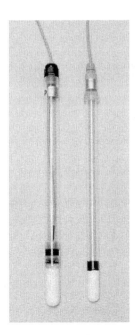

Figure 3   *Examples of tensiometers with manometer (left) and with electronic pressure transducer (right).*

Table 1   Thresholds of soil water potential (tension) and crop responses.

Scale of water potential as it reflects the need for irrigation

0 KPa	Saturated soil. If prolonged, anaerobic conditions likely.
1 – 10 KPa	Surplus water in the soil. If prolonged, the soil is not draining. Conditions favourable for disease development.
10 – 20 KPa	Plenty of water and air in the soil. Do not irrigate.
20 – 40 KPa	Adequate water in the soil. Prepare to irrigate as the readings approach the upper limit.
40 –60 KPa	Sufficient water in fine soils if the crop has a well-developed root system. In medium soils consider irrigation. In sandy soils, irrigate.
60 – 80 KPa	Readily available water is almost all used except on heavy soils. Wilting imminent.
80 –100 KPa	Do not let the soil get this dry. On re-wetting there may be problems with tuber growth cracking. Tensiometer no longer reliable.

flow of water along the tensiometer. Allow time to let the tensiometer cup reach equilibrium with the soil water before taking the first measurements.

Water balance

The rate of evaporation from a crop surface can be calculated from weather variables in several ways. Where that estimated rate of loss of water is combined with measured values for water input, rainfall and irrigation, it is possible to maintain a water balance of the crop, rather like a bank balance, and so avoid the necessity of making either direct or indirect measurements of either plant or soil water status directly. The problem then reduces to making measurements of weather and crop that are sufficient to allow calculation of the rate of water loss. Occasional measurements of soil moisture status can however improve irrigation efficiency, thus reducing water use. This in turn will reduce leaching risks of fertiliser or nitrogen losses by denitrification. Several methods are available, the best known of which is Penman's equation (Penman, 1948) which, although empirical, has the merit of being physically based. However, the parameters in that equation were chosen to estimate the rate of water loss from a well-watered grass surface, in effect as if evaporation occurs from the outer surface of the leaves. Monteith's form of the Penman equation (Monteith, 1965) introduced surface parameters and can be used to predict evaporation rates, specific to a given crop at a particular stage of development. All of these methods of computation require weather data. Formerly, those had to be obtained from one of the meteorological services but now one can install an automatic weather station at reasonable cost (Fig. 4). An advantage with such equipment is that it generally comes with a programmable data logger that can compute potential evaporation rates. Sources of advice on weather can be found on several web sites.

Suppliers of automatic weather stations include: Automatic weather station by Campbell Scientific; Dynamet by Dynamax Inc., USA; Delta-T Weather Station by Delta-T Devices; Hardi Metpole by Hardi, Denmark.

Where the grower must rely on less sophisticated means, actual evapo-transpiration is calculated from a simpler derivation, the reference crop evapotranspiration adjusted by a crop coefficient to give an estimate of actual evapotranspiration. We now present these here.

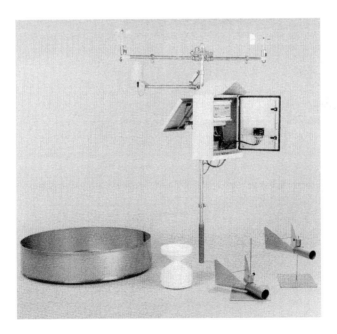

Figure 4   A weather station for recording wind direction and speed, radiation, air temperature and humidity, precipitation and pan evaporation.

Determine Et_0 – reference (potential) evapotranspiration

The weather is the most important set of factors determining the crop water requirements. Potential evapotranspiration is the rate of water loss from a well-watered crop and is measured in mm of water per day. Reference evapotranspiration (denoted Et_0) is defined as the evapotranspiration of an actively growing green grass cover, kept 8 to 15 cm tall, completely shading the ground and not short of water. Et_0 is used as a driving variable in many of the empirical models used to describe soil-plant-atmosphere processes. Daily values of Et_0 can be calculated with sufficient accuracy using any of several methods such as the *Penman equation, a Radiation equation, and Pan Evaporation.* These are summarised briefly here.

Penman method

$$Et_0 \text{ } Penman = C1[W1 \times Rn + (1\text{-}W1) \times f(U) \times (ea - ed)]$$

Where:

ea, ed = saturation vapour pressure at mean air temperature and actual vapour pressure (mbar) respectively. (ea – ed) is the 'saturation deficit', and ed is estimated as $ed = ea \times RH/100$ (where RH is the relative humidity of the air, in %).

U = wind run in km/day measured at 2 m height and $f(U) = 0.27(1 + U/100)$.

Rn = estimated total net radiation expressed in terms of equivalent water evaporated, mm/day: $Rn = 0.75Rs - Rn_{Long}$ and $Rs = (0.25 + 0.50) \times n/N)Ra$. Ra is extra-terrestrial radiation expressed in mm/day, and n/N is the ratio of actual sunshine durationto the maximum possible sunshine duration. Rn_{Long} is net long wave radiation expressed in mm/day and is a function of mean air temperature, T_{mean}, ed and sunshine duration, $Rn_{Long} = f(T) \times f(n/N) \times f(ed)$.

W1 = a temperature and altitude dependent weighting factor.

C1 = an adjustment factor for ratio U_{day}/U_{night}, for RH_{max} and for Rs.

Radiation method

Reference evapotranspiration (mm/day) can be estimated using a shorter set of variables by using the 'radiation method':

$$ET_0 = C2(W2 \times Rs)$$

Where:

Rs =   as above.

W2 =   a temperature and altitude dependent weighting factor.

C2 =   an adjustment factor made graphically on W2 × Rs using estimated values of RH_{mean} and $U_{daytime}$.

Pan Evaporation Method

In this method, reference evapotranspiration is obtained after empirical adjustments are made to the measured rate of evaporation from a standard, 'Class A' evaporation pan (fig. 4). Data required are mean pan evaporation (E_{pan} in mm/day), estimated values of mean relative humidity (RH in %) and mean wind run (U in km/day at 2 m height) and information on whether the pan is surrounded by a cropped or dry fallow area. Then:

$$Et_0 = k_{pan} \times E_{pan}$$

Where:

E_{pan} =   evaporation in mm/day from unscreened class evaporation pan

k_{pan} =   a pan coefficient

These methods are described in detail by Doorenbos & Kassam (1979). The choice of method used is governed by the meteorological data that are available.

Determine Et_c – crop evapotranspiration

The actual rate of evaporation from the crop is limited by the amount of plant available water in the soil through its effect on the surface properties of the crop. – For example, closure of stomata and wilting. A 'crop coefficient' is used to estimate actual (crop) evapotranspiration as a fraction of Et_0.

For a given climate, crop and crop development stage, the crop evapotranspiration Et_c in mm/day of the period considered is:

$$Et_c = k_c \times ET_0$$

Where:
$k_c =$ the crop coefficient
$ET_0 =$ reference evapotranspiration.

The crop coefficient is determined empirically and its value varies with the crop and its development stage, and also with wind speed and humidity. The crop coefficient increases from a low value at the time of crop emergence to a maximum value during full crop cover and then declines as the crop matures. For potatoes the value of the crop coefficient averaged over the whole growing season is between 0.75 – 0.9. However, it is better to use unique values for each crop stage (Table 2).

Table 2   *Crop coefficients for potatoes as a function of development stage*

Crop development stage (and associated % ground cover)					Total growing period
Initial (0 – 10%)	Crop development (10 – 80%)	Mid-season (805 to start of senescence)	Late season (during senescence)	At harvest	
0.4-0.5	0.7-0.8	1.05-1.2	0.85-0.95	0.7-0.75	0.75-0.9

Simulation models

Many mathematical models have been developed to simulate soil and plant processes including crop water use and soil water balance. They range in complexity, degree of validation and user friendliness but some are used in practice by farmers to guide irrigation scheduling and fertilization. Some models can be used to estimate water use throughout a growing season. Others may be used for interpolation of soil moisture status between actual measurements.

More detail about the modelling approach will be given in Chapter 5.3. Although not suited to management decisions, graphically- based models can be useful to gain an insight into the dynamics of soil water movement, infiltration, transpiration, runoff, through drainage, and crop growth. An example of this is WaterMod 2 (Greenhat Software).

Table 3   *Appreciation of the suitability of several techniques for determination of plant and soil water for various purposes. Appreciation includes cost, ease of installation, operation and automation, and accuracy (Scale: # not suitable, *** highly usable).*

Technique	Farmers	Consultancy agencies	Research	Overall
Plant:				
Pressure chamber	#	#	***	**
Psychrometer	#	#	**	*
Relative Water Content	#	*	***	**
Stem flow	?	*	***	*
Porometry	#	*	***	*
Infrared thermometry	**	**	***	**
Soil:				
Gravimetric	#	#	**	*
Neutron probe	#	**	***	**
Gamma probe	#	#	*	#
TDR	**	***	***	***
FDR	***	***	***	***
Gypsum block	#	#	#	#
Psychrometer	#	#	*	#
Tensiometer -				**
Vacuum gauge	*	*	#	
Pressure transducer	**	**	**	
Water balance	***	***	***	***

Conclusions

Some of the techniques for measuring soil water status discussed here are only suitable for research purposes, others can be used by consultancy agencies and some are used by farmers for practical everyday use (see Table 3). Comments on suitability have been made within the text and are summarised here.

The most widely used technique for scheduling irrigation is maintaining a water balance, coupled with calculation of evapotranspiration. Among the instruments available, the tensiometer was the most commonly chosen until recently. The recent introduction of capacitance-based probes, here grouped under FDR, has changed this and these instruments are so much better for field use than those previously available that they are rapidly supplanting them. However, even the capacitance probes still present a sampling problem and so calculation of the water balance will still be the favoured tool for irrigation scheduling for many users. Where site-specific irrigation can be employed, a case can be made for mounting infrared thermometers on the irrigation boom.

Recommendations

1. The potato crop can rarely be grown at its best without the possibility of applying irrigation.
2. Irrigation should not be applied without proper scheduling
3. The most appropriate form of scheduling, whether calculated directly or obtained as a service from a consultant is to calculate a water balance
4. Where item 2 above is not available or where checks are to be made directly in the crop, the most appropriate approach is to measure soil water. Modern capacitance-based probes are probably the best suited to field use.
5. Any method chosen for direct measurement should be used with an adequate sampling protocol.

4.3   Practical use of soil water measurement in potato production

R.J. Bailey

What are the effects of water stress on potato crops? How can irrigation be used to minimise these effects? How can irrigation be planned so as to get maximum benefit from the water, and eliminate unnecessary wastage? How can soil moisture measurement assist with the planning process? Are some methods of measurement better than others? How frequently do measurements need to be taken? How many measurements are needed on each occasion? Where in the field should measurements be taken?

Introduction: Influence of water on quality and yield

Potatoes are more sensitive to water stress than other commonly grown crops. There are two reasons for this: (i) even moderately compacted soil will impede potato root growth, and so potatoes often form shallow and inefficient rooting systems (ii) even mild water deficits will cause potato leaf stomata to close, and this decreases growth.

This sensitivity to water stress is common throughout Europe. For example, in inland areas of Portugal, crops without irrigation usually yield only 9-12 t/ha, but irrigated crops regularly achieve yields of 30-50 t/ha. Similarly, on the drier eastern side of the UK, yields without irrigation are sometimes as low as 20 t/ha, and irrigated yields regularly attain 50-55 t/ha.

In regions of low rainfall, water stress can only be relieved by irrigation. It is not possible to overcome the effects of water stress by increasing nitrogen applications. Droughted crops will not respond even to normal levels of fertiliser, and unirrigated potatoes in low rainfall areas should receive less nitrogen fertiliser than irrigated crops, in accord with their lower yield expectation.

Water stress also affects tuber quality. The visual appearance of tubers with 'common scab disease', caused by *Streptomyces scabies*, renders them unacceptable to the consumer. This disease is worse in dry soil conditions, and even slight moisture deficits soon after tuber initiation can result in high levels of common scab.

Tuber dry matter content and specific gravity are important components of processing quality. Although largely determined by cultivar choice and sunshine levels, they can be influenced by drought stress. This relationship is confusing, however. While UK experiments generally demonstrate higher dry matter associated with drought stress, experiments in the United States suggest the opposite (Miller & Martin, 1987; Wright & Stark, 1990).

Another manifestation of water stress can sometimes occur during tuber development, resulting in different proportions of starch and sugar along the length of the tuber. In a severe form of this disorder, one end of the tuber appears translucent. Such tubers show variable colour change upon frying, making them unsuitable for the processing industry. Growth cracking and secondary growth protuberances can also be caused by short periods of water stress.

Detailed discussion of the effects of water stress on potatoes, and the benefits of irrigation, can be found in several reviews (Wright & Stark, 1990; Bailey, 1990; Harris, 1978; Van Loon, 1981).

Water is thus important for quality and yield of potatoes, and many growers irrigate to guarantee a plentiful water supply. The best results from irrigation are obtained from careful planning, usually referred to as 'irrigation scheduling'. This section explains the principles of irrigation scheduling, and shows how soil moisture measurement can be used in this process. It also explains why some methods are better than others and gives advice on the frequency and best location for measurements. This should enable farmers to make the best use of their irrigation facilities.

Principles of irrigation scheduling

After heavy rain or irrigation, the surface layers of a free-draining soil are temporarily saturated, and all the pores of the soil are filled with water. This is usually a short-lived situation, and surplus water soon starts to drain downwards, wetting the layers below. Drainage continues until each soil layer is holding the maximum quantity of water that it can retain against the pull of gravity. The force holding this water is surface tension, and a soil in this state is said to be at field capacity. It is convenient to consider this water in terms of the suction required to remove some of it. At field capacity, water removal requires a suction somewhere between -5 kPa and -10 kPa, and we call this value the soil moisture tension. Although not strictly correct, for most practical purposes we can assume that light soils drain to field capacity throughout the rooting zone within a couple of days after being saturated.

Heavy clay soils are not free-draining, and may take much longer to reach field capacity. It is worth noting that while the concept of field capacity is convenient for practical use, entire soil profiles at field capacity rarely exist, especially for slow draining soils; by the time the lower levels have drained to field capacity, the upper levels have often dried beyond this point as a result of surface evaporation or crop water uptake. Fortunately, because potatoes are a shallow rooting crop, we are generally concerned only with the upper 70cm of soil and errors arising from this effect are small.

As a crop removes water from the soil, the soil is described as having a certain soil moisture deficit (SMD). The SMD at field capacity is zero (see Fig.1). If a crop extracts 50 mm of water from a soil, and there has been no rainfall or irrigation to replenish it, the soil is said to have an SMD of 50 mm. If, during the following four days, the crop extracts a further 10 mm of water, the SMD is then 60 mm. If rainfall or irrigation then adds 20 mm of water back to the soil, the SMD becomes 40 mm. As this process occurs, and water is removed from the soil, the water remaining is that which is held at a greater tension i.e. it requires a greater suction to remove it.

At field capacity, or when the SMD is low, water supply is unrestricted and crop water uptake is at a maximum, the rate being determined by weather the SMD

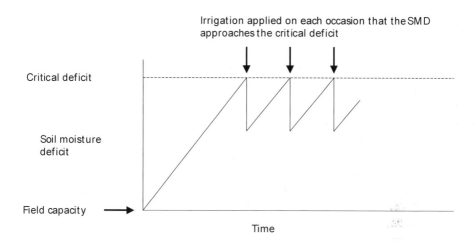

Figure 1   Diagrammatic representation of irrigation scheduling.

increases, the soil water tension increases, water becomes more difficult to extract, and plants respond by closing stomata and thereby reducing water loss to the atmosphere. This is an important stage, and the SMD at which it occurs is known as the critical SMD, because closing the stomata will result in less intake of carbon dioxide, and therefore lower rates of photosynthesis and potentially lower yields. The soil water tension at this stage is known as the critical tension. We can also think of this stage as the refill point, because it is the point at which we must irrigate to refill the soil profile with water if we are to avoid any yield loss.

Irrigation scheduling should aim at maintaining soil moisture somewhere between the refill point and field capacity. This is achieved by switching irrigation on before the soil dries to the refill point, and ensuring that the amount applied does not exceed the quantity required to return to field capacity. Irrigating beyond field capacity will only initiate drainage, reduce irrigation efficiency, and possibly give rise to problems associated with poor soil aeration. An exception to this rule applies when using saline water for irrigation, but this will be discussed later. A further sophistication is employed by some growers, in that they do not irrigate fully to field capacity when there is the possibility of rain, but leave a margin to allow for the rain to be fully absorbed by the soil and benefit the crop.

Maintaining soil moisture

In arid regions, where rainfall is insignificant during the growing season, it may be possible to design irrigation plans based on long-term average climatic data. For each specific soil type and planting date combination, growers can obtain a schedule of irrigations for the season. Throughout much of Europe, however, rainfall represents a significant contribution to crop water needs. Rainfall is often localised, of varying intensity, and impossible to predict accurately in advance. The only way to take this into account is by using irrigation schedules that involve real-time plant measurements, soil water measurements, or rainfall measurement integrated into a water balance calculation. These have all been discussed in detail in the previous section. Here we shall concentrate on some of the practical issues associated with soil water measurement.

With all methods of soil water measurement, there are four questions:
- how many measuring points do I need in each field?
- where in the field should the measuring points be located?
- how do I determine the field capacity and refill points for my soils?
- how often do I need to take measurements?

The answers to some of these questions depend on which soil water measurement technique is used, but there are some common guidelines. We shall consider these first and then discuss individual techniques.

Avoid field boundaries. Field boundaries are not usually representative of the field, and measurement points should be at least 20 metres away from the field boundary.

Avoid run-off intercepts. Irrigation or heavy rain will often cause surface run-off which can collect in hollows or vehicle wheelings. These will not be representative of the whole field.

Avoid areas of poor crop growth. Diseased or poorly-growing patches of crop may use less water than the surrounding healthy crop.

Take the measurement from within the plant rows, mid-way between normally spaced plants.

In most situations, a field will be a single irrigated unit. If the field is split between crops, or planting dates, which are to be irrigated differently, then each sub-unit must be monitored separately.

Some irrigation techniques have specialised requirements. For surface irrigation, monitors should be sited approximately mid-way from the head of the field. Depth of measurement is discussed below, but centre-pivots and trickle irrigation are often used to apply small quantities of water every day. In these circumstances, the wetting and drying cycle of the soil will be confined to a narrow surface layer, and the depth of measurement should correspond with this.

How often are measurements required? A measurement is required whenever the grower is unsure of the soil moisture level. In the absence of any experience, a newcomer to soil moisture measurement will need to take daily readings. As experience is gained, growers will soon learn how quickly their soils dry and measurements can be taken less frequently. Growers employing a commercial service, such as a neutron probe contractor, will have the immediate benefit of the contractor's experience, and may only need readings on a weekly basis.

We shall now discuss individual techniques in detail.

Manual examination

Many growers instinctively believe that the best approach consists of digging a spadeful of soil, and manually examining the colour and 'feel'. As a general rule, growers should not rely on this. It can be successful, but note that (i) the soil sample must be taken from the whole rooting profile, i.e. the top 50-60 cm. (ii) one or two sampling locations will not suffice. Indeed, colour and 'feel' of soil are usually variable within a field, and at least six sampling locations are needed on each occasion.

Being both onerous and imprecise, this technique is best avoided in favour of the other techniques available.

Soil moisture tension

Soil moisture tension is usually measured with tensiometers. These should be sited in pairs, the upper partner 30 cm below the top of the ridge, and the lower partner at a depth of 50-60 cm. Irrigation is started when the upper tensiometer indicates that the soil is at the critical soil moisture tension or refill point; the critical tension is approximately -30 kPa for sandy soils, -40 kPa for sandy loams, -55 kPa for silt loams, and -65 kPa for silty clay loams (Wright & Stark, 1990). Growers with knowledge of local transpiration rates can make fine-tuning adjustment to these numbers. For transpiration rates outside the range 2.5-5 mm per day, adjust the refill point by 10 kPa. Thus on sandy soils, if the average transpiration is only 2 mm per day, the critical tension is -40 kPa, but where average transpiration is 6 mm per day the critical tension is -20 kPa. If irrigating to prevent common scab disease on light soils, a refill point of -20 kPa should be used for six weeks after tuber initiation.

As irrigation is applied, the upper soil layers become wetted and the upper tensiometer will register a marked decrease in tension, returning to near field capacity. Only when the whole profile is fully wetted will the lower tensiometer register a tension decrease, and this is the signal that sufficient irrigation, or even too much, has been applied. Growers using tensiometers for the first time are advised to stop when water reaches the upper tensiometer. Continued observation of the lower tensiometer will soon indicate whether sufficient water has been applied. With continued use, growers will soon learn how much water is required before this occurs.

How many pairs of tensiometers are required per field? Three pairs should be used whenever possible. Tensiometers sometimes give bad readings owing to poor conductance between the soil and the ceramic cup, and three pairs allows erroneous readings to be identified easily and discarded.

Where there is known soil variation across the field, irrigation can become complex. If it is not possible to split the field into sub-units, but the whole field is to be irrigated as a single unit, a compromise must be reached. In most cases, growers will wish to irrigate according to the requirement of the sandier areas, even though this may involve irrigating the rest of the field more often than is strictly necessary. In such cases, the tensiometers should be placed in the sandier areas.

Soil moisture content - fixed sampling points

Soil moisture content is greatly affected by even small variations in soil texture, and so we are likely to find a great deal of variability within a field. To establish an average soil moisture content for a whole field with neutron probes will require many measurement points. When scheduling irrigation with neutron probes, we overcome this problem by (a) returning to the exact same location each time and (b) measuring the change in soil moisture content. Table 1 illustrates this point, using actual data from field measurements with neutron probes at ADAS Gleadthorpe Research Centre, UK.

On the first measurement date, the three locations within the field exhibit soil moisture contents that vary widely, with a range of 21 mm, and it is not possible to estimate an average for the whole field with any degree of confidence from these three points only. The same is true of the second measurement date. The change in soil moisture content shows much less variation, however, with a range of only 3 mm (i.e. -8 to -11 mm). For practical irrigation purposes, we can conclude that the crop consumed approximately 10 mm water during this 4-day period.

Normal practice involves taking a reading at field capacity (two to three days after saturation), when the soil moisture deficit (SMD) is assumed to be zero. Subsequent readings are then related back to this first measurement by

Table 1   Soils moisture content (mm) in upper 70 cm soil

	A	B	C
Day 1	99	120	109
Day 5	91	109	99
Change	-8	-11	-10

Table 2   *Critical soil moisture deficit or refill point for maincrop potatoes according to soil texture.*

Soil texture	Critical SMD (mm)
Sand	30
Loamy medium sand over sand	35
Loamy fine sand over sand	40
Medium sandy loam, sandy clay, sandy clay loam, clay or silty clay	55
Clay loam, silty clay loam, fine sandy loam, sandy silt loam	60
Silt loam	75

calculating a difference, which represents the SMD at the time of each subsequent measurement. The critical SMD or refill point for each soil type, based on a rooting depth of 70cm, is given in Table 2. It should be remembered that these critical deficits or refill points represent the driest condition that can be allowed before significant yield is lost. It is not necessary to wait until reaching this condition before irrigating; irrigation can be applied earlier provided sufficient margin is left to make good use of any subsequent rainfall that might occur. It can be a good tactic, even on the heavier soils, to irrigate whenever the SMD reaches 35-45mm, but ensuring that irrigation is stopped early enough towards the end of the season so that SMDs rise to the levels in Table 2 before harvest.

Where local transpiration rates are known to be less than 2.5 mm per day, these critical deficits can be increased by 10 mm. Similarly, where transpiration rates are in excess of 5 mm per day, they should be decreased by 10 mm. When irrigating to prevent common scab disease on light soils, a critical deficit of 15 mm should be used for six weeks after tuber initiation.

Measurements should be taken throughout the upper 70 cm of soil. Two sampling points per field is the bare minimum necessary, but three is much preferred. Where a field contains soil of varying texture, growers should select a critical deficit from Table 2 that corresponds with the lighter soil but, unlike tensiometers, neutron probe measurement points should be sited in the heavier textured parts of the field.

Soil moisture content - different sampling points

Some soil moisture measurement techniques (e.g. gravimetric analysis, see previous section) necessitate removal of the soil. It is then not possible to return to the exact same location each time we take a measurement, but we are forced to use a different soil sample on each occasion. With such methods, we must obtain an accurate measure of soil moisture throughout the whole field every time we take a measurement, and this requires at least 20 samples per field. For most growers, this is too arduous a task, and the technique is not recommended.

Apart from the number of sampling points, the technique is similar to that previously described for neutron probes. Samples are taken throughout a 70 cm profile when the soil is at field capacity, and subsequent measurements are compared to these in order to estimate the difference i.e. the SMD. The critical deficits in Table 2 can again be used to establish the refill point.

Soil water balance calculations

These are described in the previous section. Strictly speaking, they are not a soil measurement at all, but are a means of calculating SMD based on measured weather data. This is the single most reliable technique for irrigation scheduling. Once again, the critical deficits in Table 2 can be used with this method also.

Other techniques for soil water measurement

All techniques measure either soil moisture tension or soil moisture content, and the above guidelines can be applied. For example, electrical resistance blocks measure soil water tension and can be used as described for tensiometers. Similarly, capacitance meters and time-domain-reflectometry (TDR) meters measure soil moisture content and can be used as described for neutron probes.

Forecasting ahead

All of these techniques can be used to determine whether irrigation is required or not i.e. has the soil reached the refill point? With practice, users will be able to improve on this and forecast ahead. They will soon become familiar with the rate of change of soil moisture over several days and will be able to answer the question 'if no rainfall occurs how many days before irrigation is needed again?'.

Soil water balance calculation vs. soil water measurement

Which of these two approaches is the best? Soil water balances are easy to calculate, lend themselves to forecasting ahead several days, but are not always accurate. Soil water measurements can be very accurate, but not representative of the field unless many measurements are taken, and this is very time-consuming. A good compromise consists of using a water balance calculation, and occasionally taking measurements to correct any accumulated errors in the balance sheet. This combination combines ease of use and accuracy.

Water Quality

Various substances can find their way into water supplies and be harmful to crops if the water is used for irrigation. These include boron, calcium, iron, sodium salts and some organic contaminants. The element which most frequently limits the suitability of water for irrigation is chloride, usually derived from dissolved sodium chloride (common salt).

Saline water contains both sodium and chloride. Problems are most commonly encountered in coastal areas where sea water finds its way into ground water or surface supplies. Although the sodium can cause damage to soil structure in clay soils, it is the problems associated with the chloride content of saline water which are of greater concern. Chloride can damage plants in two ways: by directly scorching the leaves and by inhibition of root activity.

Direct leaf scorch is usually associated with bright, sunny conditions, and if saline water has to be used the risk can be reduced by irrigating during dull periods or in the evening. It is difficult to be precise about the levels of chloride associated with such damage, because even an initially low concentration of chloride in irrigation water can soon increase to a high concentration as the droplets on the foliage evaporate and decrease in size.

It may be useful to note that in two UK experiments, irrigating with water containing 2000 mg/l chloride produced significant leaf scorch, but the resulting yields were higher than unirrigated potatoes.

In regions of higher winter rainfall, accumulation of chloride in the soil is prevented by winter leaching. In the absence of any rainfall, however, the salts accumulate and eventually slow water uptake, producing effects that are similar to drought stress. The only practical measure to overcome this involves excessive irrigation beyond field capacity, in order to leach the salts below the rooting profile. The amount of excess water required, termed the leaching fraction, depends on the salinity level, so it is not possible to give precise guidelines here. Growers with this problem should seek local advice.

Conclusions

Potato yield and tuber quality are particularly vulnerable to drought stress. This can be alleviated by irrigation, but careful planning or scheduling is required if we are to obtain the maximum benefit from irrigation without undue waste of water. Among the many methods of irrigation scheduling, maintaining a water balance using weather data is the best. However, soil moisture measurements can also play a useful role in irrigation planning, provided they are taken carefully, paying due regard to the number of sampling points, location of sampling points, frequency of measurement etc. Indeed a combination of measurement and water balance computation is the ideal method.

Recommendations

1. Plan or schedule irrigation applications to obtain maximum benefit.
2. If using soil moisture measurement, ensure a sufficient number of measurements are taken.
3. Avoid taking measurements from unrepresentative locations such as field boundaries, sheltered hollows etc.

4. Practice makes perfect. Keep accurate records, constantly refer to them, and learn from experience.

5 Nitrogen and water – decision support systems

5.1 Nitrogen decision support systems in potato production

G. Hofman & J. Salomez

Nitrogen decision support systems can deal with different aspects related to nitrogen and factors influencing these. Depending on the complexity of the system more or less information is necessary. Applying a single dressing or not? If not, does top dressing depend on developmental stage? What kinds of expedients do we have to decide when to apply supplementary N dressings? What kind of manure to apply, advantages and disadvantages? When and how to fertilise and how accurate can we fertilise? How to improve on existing N decision support systems? Yield maximisation, yield optimisation, how to account with environmental legislative aspects?

Introduction

The efficient use of fertilisers, both organic and mineral, is of paramount importance from an economical point of view as well as from environmental concern. However, the formulation of a correct nitrogen fertilisation advice is difficult because the optimum N application rate is situated in a narrow range and nitrogen is subjected to various processes in the soil. Figure 1 shows the different factors and processes which influence the mineral nitrogen 'pool'. This pool, associated to N-fertiliser supply, has, up to a certain level, a positive effect on yield. On the other hand, quality characteristics of the tubers do not always respond positively on the amount of available nitrogen. Systems that thus estimate the mineral N pool more exactly will contribute to a better N-management in potato crop production.

In this chapter, the accuracy and use of different nitrogen fertilisation advice systems, i.e. fixed N-rates, N_{min} method sensu stricto, N-index method and the nitrogen balance sheet method, are first reviewed. The use of simulation

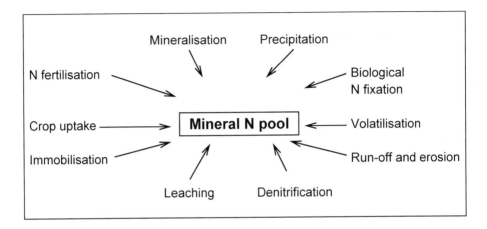

Figure 1   Factors and parameters influencing the mineral nitrogen pool.

models will be briefly discussed along with the possibilities of plant analysis to optimise N-fertilisation. Aiming to maximise N use efficiency, recommendations are also given on type, time and mode of application of nitrogen fertilisers.

Nitrogen fertilisation advice systems

Until the seventies, results of field trials with various N-levels over different years were used to identify the optimum N level for a certain crop in a region. This approach was rather inadequate because especially the (potentially) available nitrogen in the rooting zone of the crop was unknown. Further, because the nitrogen required is related to level of production, the 'answer' differed each year. Rapid and accurate determination of the mineral nitrogen in the soil profile has led to scientifically based nitrogen advice systems.

Table 1   *ADAS N recommendation system for potatoes (kg N ha⁻¹).*

	Index		
	0	1	2
Maincrop and second earlies			
Sandy and shallow soils	240	200	130
Other mineral soils	220	160	100
Organic soils	180	130	80
Peaty soils	130	90	50
Earlies/canning/seed			
Sandy and shallow soils	180	130	80
Other mineral soils	160	110	70

Fixed N-rates

The most simple type of fertiliser recommendation consists of recommending a fixed rate for the crop in all situations, regardless of soil type, field characteristics, cultivar,... Though easy and without costs for soil analysis, this method is completely inadequate.

A refinement of this method is the ADAS nitrogen index method (Table 1) (Anonymous, 1994). On the basis of past management practices and of the previous grown crop, fields are adjudged an index, ranging from 0 (low amounts of N_{min} expected) to 2 (high amounts of N_{min} expected), which give an indication of expected N_{min}-residues, but the exact N_{min} amount is unknown. The recommended N rate further depends on soil type and the organic matter content of the soil.

Both types of methods are only to be used in situations where soil sampling is not possible due to the presence of stones and in situations where N_{min} at the start of the growing period is not likely to fluctuate among fields and years. In all other situations a method which includes soil analysis is recommended (Neeteson, 1995).

Variable N-rates

N$_{min}$ method sensu stricto

The results of Van der Paauw (1963) and others, concerning the effect of residual nitrogen, are chiefly responsible for the investigations on inorganic nitrogen in the soil profile. Later on, research in different countries led to an N fertilisation advice based upon the linear relationship between the N$_{min}$ in the rooting zone of the crop at the start of the growing period and the optimum N fertilisation (Fig. 2). This method is still used in several parts of Germany and in The Netherlands.
Table 2 gives an overview of the current Dutch N-fertiliser recommendations, in function of soil type, whereby a and b represent the regression coefficients of the linear relationships.

From Figure 2, it can be noticed that, although the linear regression is significant, there is still a large variation around the calculated regression. To reduce this variation, other systems with more factors involved were introduced (see further).

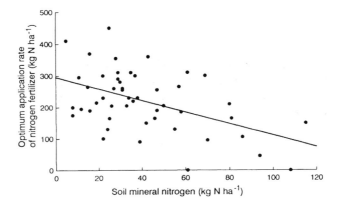

Figure 2   *Relationship (N-appl. = 300-1.8*N$_{min}$) between the amount of mineral nitrogen in the 0-30 cm soil layer at the end of the winter period and the economically optimum application rate of nitrogen fertiliser for potatoes (Solanum tuberosum L.) on sandy soils in The Netherlands (Neeteson et al., 1984).*

Table 2   *Current Dutch N fertiliser recommendations (N_{rec}) for potatoes (Anonymous, 1998).*

	$N_{rec} = a - b.N_{min}$		Sampling depth for N_{min}
	a	b	(cm)
Ware potatoes			
Clay and loam soils	285	1.1	0-60
Sandy soils	300	1.8	0-30
Starch potatoes	275	1.8	0-30
Seed potatoes	140	0.6	0-60

In Germany, the 'Landwirtschaftskammer Hannover' proposed a 'Sollwert' of 160 kg N ha^{-1}, which is much lower than in The Netherlands. The 'Sollwert' is the maximum mineral N-amount that has to be available at the start of the growing period. Positive or negative corrections are foreseen in function of soil texture, preceding crop and prolonged application of animal manure (Baumgärtel, 1997).

N-index method

The Pedological Service of Belgium proposed the N-index method in the early 80's (Boon, 1981). Besides the N_{min} amount, other factors, up to a maximum of 18, are also included into the N-index system. Depending on the history of the field, one or more of these factors can be omitted.

$$\text{N-index} = X_1 + X_2 + X_3 + \dots + X_{16} + X_{17} + X_{18}$$

whereby:
X_n : the various factors

These factors can be divided into 3 groups (Vandendriessche et al., 1992):

N_{min} (X_1):
Being the mineral N in the soil profile until a depth of 60 cm;

Mineralisation (X_2-X_9):
Being the factors responsible for the N release from soil organic matter and all kind of incorporated material, e.g. green manure, crop residues, animal manure, compost,...

Negative factors (X_{10}-X_{18}):
Being factors which have a negative effect on the N availability, e.g. compaction, too low pH, possible N leaching,...

The optimum N fertilisation will thus be calculated as follows:

N advice = a - b.N-index

whereby
a, b   : indices, depending on cultivar and destination of the tubers (Table 3).

Calculation of this linear relationship between the N-index and the optimum N fertilisation results in a more significant relationship than showed in Figure 2.

Table 3   N fertiliser advice depending on tuber destination (same N-index) (Bries et al., 1995).

	N advice (kg N ha^{-1})
Early potatoes (Première)	100 - 150
Ware potatoes (Nicola)	80 - 130
Ware potatoes (Bintje)	150 - 200
Fries (Bintje)	120 - 170
Chips (Saturna)	130 - 180
Starch potatoes (Prevalent)	120 - 170
Seed potatoes (Bintje)	50 - 100

Nitrogen balance sheet method

This method was first developed in France and the USA (Hébert, 1973; Carter *et al.*, 1974), but is with some minor changes also used in Belgium and The Netherlands (Hofman, 1983; Neeteson *et al.*, 1988).

The theoretical N fertilisation is calculated as follows:

N need of the crop + Residual N_{min} in the soil profile at harvest*	=	N_{min} before planting + N mineralisation + N fertilisation

*   The residual N_{min} in the soil profile 0-60 cm at harvest is the amount of mineral N which remains in the rooting zone at optimum N fertilisation, potatoes normally leaving values ranging from 40 to 100 kg N ha^{-1}.

The practical N fertilisation depends on the expected losses. These are estimated to range between 5 and 20%, mostly dependent on soil texture.

The above mentioned methods do not take into account (fixed rate and N_{min} method) or only estimate (N-index method and N balance sheet method) the amount of nitrogen which is to be mineralised from soil organic matter during crop growth. To better cope with post-planting mineralisation, other methods and expedients are in use, all referring whether or not to use an additional top dressing. Other reasons to use a top dressing are possible improvement in fertiliser N-use efficiency, possible yield increase, improvement of quality and the decrease of the environmental impact of N fertilisation.

KNS-system

The 'Kulturbegleitenden N_{min} Sollwerte-System' (KNS-system), introduced in Germany by Lorenz *et al.* (1985) and taken over in The Netherlands as the 'N-BijmestSysteem' (NBS-system) (Breimer, 1989) was first developed as an aid towards the N fertilisation of vegetables. It involves the measurement of the residual N_{min} at fixed intervals during the growing period and the comparison with target values (Pannier *et al.*, 1996). Though promising, the method does not seem really suitable for potatoes as a second N-dressing in most cases does not show real advantages (see further) and a correct sampling for N_{min} determinations during the growing period is difficult (Hofman *et al.*, 1993). However, splitting the N-dose gains importance in the UK and The Netherlands. The advantage of this system is that you can adapt your supplementary N-dose according to N-mineralisation and the performance of the crop in the early stage of growing. When irrigation possibilities are provided, splitting the N-dose becomes even more important. The interest of the application of supplemental N-dressings during the growing season is discussed into subchapter 5.2.

Simulation models

With simulation models it is possible to calculate, on a daily basis, the availability of nitrogen to the crop and the nitrogen uptake and the growth of a crop, using average or actual weather data and soil, crop and field parameters as inputs. Simulation models can thus be used to estimate the fertiliser nitrogen requirements of a crop at any time during the growing period but also the environmental side effects of nitrogen fertiliser application can be estimated. In order to keep these models as simple as possible and to keep the number of parameters and input data at a minimum, they will have to be simplified as much as is justified by the soil, crop and climatic conditions in a given environment (Neeteson, 1995). The main disadvantages of (simplified) models is still their need for a numerous amount of data, which are not always readily available and extrapolation is sometimes difficult as they were mostly developed for a specific region or country. The advantages and disadvantages of using simulation models is further discussed into the modelling chapter.

Plant analysis (Petiole sap analysis, Chlorophyll-meter readings,...)

Plant analysis is used to check the nitrogen status of a crop during the growth period. The idea behind plant analysis is that the crops themselves are the best indicators of the supply of nitrogen by the soil as well as the nitrogen demand of the crop and its abilities to absorb the nitrogen available in the soil. When the nitrogen status appears to be too low, it can be decided to apply additional fertiliser nitrogen. Plant analysis methods have the advantage that a second N fertilisation can be delayed and the mineral N supply from soil organic matter can be at least partly introduced in the advice system. However the 'translation' of values obtained into amounts of fertiliser nitrogen to be applied to compensate for the nitrogen deficiency is up till now very difficult and optimal timing for second N fertilisation to be applied is not easy to define. The use of plant analysis in optimising the potato N fertilisation is largely discussed in chapters 2.1 to 2.5 of this handbook and quantitative information about existing systems can be found there.

Type of fertilisers

Organic fertilisers

The positive influence of farmyard manure (FYM) on potato production is recognised since a long time. It ameliorates soil structure, soil aeration, water holding capacity of the soil,... However, intensive dairy, pig and chicken production resulted in the production of slurry, with a much lower dry matter (DM) and organic matter (OM) content (Table 4). The positive influence on soil structure will therefore be more limited compared to FYM. Other organic materials like compost, but also industrial organic waste products are presented to agriculture as fertilisers or at least as soil ameliorates.

To maintain the OM content in the plough layer at a certain level and to keep soil structure at optimal level, the application of organic material has to be encouraged. On the other hand, the use of organic material complicates a correct N fertilisation because the N restitution from this material is still not well known and has to be estimated. High soil mineral N residues at harvest of the potato crop are often caused by a miscalculation of the mineralised nitrogen from incorporated organic material. In such situations and previous to application of a second N dressing, the use of plant analysis as N status indicator can help to better match N supply to N demand.

Table 4   Composition of some organic manures and some widely used mineral fertilisers (Bries et al., 1995; Anonymous, 1998).

Source	Dry matter	Organic matter	Total N
Slurry (kg (1000 l)$^{-1}$)			
Cow	95	60	4.4
Pig	85	60	6.5
Chicken	160	90	9.0
Farmyard manure (kg (1000 kg)$^{-1}$)			
Cow	240	140	5.5
Pig	230	160	7.5
Chicken (dry)	600	420	24.0
Mineral fertilisers (%)			
NH_4NO_3	-	-	27
$(NH_4)_2SO_4$	-	-	26
Urea	-	-	46

Mineral N fertilisers

The three main mineral N-fertilisers in potato crop production are ammoniumnitrate (NH_4NO_3), ammoniumsulphate (($NH_4)_2SO_4$) and urea ($CO(NH_2)_2$) (Table 4). The most common used fertiliser is NH_4NO_3, although $(NH_4)_2SO_4$ is often used because potato growth is favoured in a light acid environment. Besides as a solid fertiliser, urea is used in small amounts as a liquid fertiliser for foliar applications. The use of a remainder supply of urea-N, i.e. 50 kg N ha^{-1} split applied at a 7 to 10 days interval, from complete emergence the 15th of August on, enhances tuber yield and size. Such effects seem to be particularly relevant in situations with lower soil mineral N-supply such as drought stress (Goffart & Guiot, 1996; Goffart et al., 1999).

The use of slow release N-fertilisers can be considered in situations where leaching losses are likely to occur, e.g. shallow sandy soils, but a supplementary N dressing in most cases can do as well.

Application of nitrogen fertilisers

Time of application

Nutrients such as nitrogen, which are subjected to easy losses, have to be theoretically applied concomitant with the N-uptake by the crop. However, traditionally the N-fertilisation for potatoes is mostly applied before planting. Main reasons are the additional costs and the rather low N-efficiency of a second N-dressing, although efficiency is also very dependent on the form and mode of application of the N fertiliser.

N-fertilisation of more than 200 kg N ha^{-1} at once can result in salt damages and therefore a split application will be recommended. Generally, it is recommended to apply a second N-dressing the latest at tuber initiation to avoid risk of poor use of solid N fertiliser due to dry weather conditions. On the other hand, plant N-uptake should not be hampered while using foliar N fertiliser or fertigation later in the season. However, it is recommended to avoid too late N fertiliser application which could be of poor interest for tuber yield and negatively affect tuber quality (dry matter content, reducing sugars content, nitrate content), especially for late cultivars, and finally also increase soil mineral N residues at harvest.

Mode of application

Because of the restricted rooting depth and the limited root distribution, the N-efficiency in potato cultivation is low.

Generally, N-fertilisers are broadcasted shortly before planting. By earthing up, fertiliser nitrogen is brought into the ridge and at least to a certain extent above the tuber and thus hardly available. An application in the row at about 5 cm below the plant tuber and at 7 to 10 cm from the planting line has some potential advantages (Demeyer, 1993; Hofman *et al.*, 1993):

- a better availability of mineral N;
- a reduction of N-leaching losses due to a retardation of the nitrification process in the early stage of the growing period;

- the reduction of ammonia volatilisation after the application of urea or ammonia-rich fertilisers on calcareous soils.

Although theoretically advantages could be expected, results of several years experiments did not show real significant differences between broadcast and band application, except for a better grading of the tubers (Salomez *et al.*, 1997). More significant benefits can be expected on soils with low soil fertility.

If an N-shortage is recognised later in the growing season (see chapter 2 for use of plant N status indicators), N foliar application can remediate it to a certain extent. The application of liquid foliar urea comparatively to solid ammonium-nitrate as top dressing can improve N-use and N-efficiency (yield and tuber size essentially) (Goffart & Guiot, 1996; Goffart *et al.*, 1999), without additional cost as foliar-urea application can be combined with fungicide treatments.

Possibilities of fertigation will be discussed into chapter 5.2.

Legislative constraints

To make agriculture sustainable and so to restrict the harmful effects of nitrogen application, the European Union has set up some legislation, known as the Nitrate Directive (Anonymous, 1991). The main purpose of the Directive is the limitation of nitrate in ground- and surface waters to 50 mg NO_3^- L^{-1}. Having an average drainage surplus of 300 L $(m^2$ $year)^{-1}$ and accepting a denitrification loss of 50%, the above-mentioned limit of 50 mg NO_3^- L^{-1} is thus obtained when ± 70 kg N ha^{-1} drains out of the soil profile.
Therefore, action programmes like designating 'Vulnerable Zones', and establishing a 'Code of Good Agricultural Practice' have to be introduced.

These measures can have a (major) impact on potato production as the application of animal manure should not exceed 170 kg N ha^{-1} in vulnerable zones, possibly leading to a lower physical soil fertility when no other steps are taken, e.g. the use of green manures.

Neeteson and Wadman (1991) showed that a yield reduction of 5-10%, mostly tubers >50 mm, is likely to occur when the recommended rates have to be adjusted as to obtain at harvest an acceptable residual N_{min} amount. On the other hand, reducing the recommended rate by about 25% did in most cases not affect yield and the chance of trespassing the limit of 50 mg NO_3^- L^{-1} was almost nil (Neeteson, 1989b).

Conclusions

Various types of nitrogen decision support systems have been presented. Traditional methods of N fertiliser recommendations include the method of applying fixed rates in all situations and the N_{min} method. They are only satisfactory in the sense that they allow the farmer to achieve maximum yields, but they make no correction for mineralisation out of soil organic matter and organic materials and hence possible losses to the environment can be high. On the contrary, the N-index system and the balance sheet method explicitly take into account an estimated N mineralisation, thus better matching N need and N fertilisation and aiming at optimise yield. As further refinements and expedients, the KNS system, plant analysis and simulation models are at our disposal.

Potatoes benefit from the application of organic materials, mainly through their action on soil physical properties. Potato growth is favoured in a light acid environment, but all sources of mineral N fertilisers can be applied. The mode of application, both broadcast and band application, in most cases does not show differences in yield, but band application can cause a shift towards a higher grading class. If legislative constraints do impose yield losses of up to 10% (mainly tubers >50mm) band application or foliar urea application seem promising tools to overcome these losses.

Recommendations

1. Main rule is to ask for N-advice preferably with a system including at least a measurement of soil mineral nitrogen availability at planting time and an estimation of N mineralisation in the course of the growing season, especially in situations with frequent organic matter return (FYM, slurry).
2. N-use efficiency can be improved with band or foliar application of N.
3. There is a general trend for recommendation of splitting of nitrogen advice, combined with plant analysis as N status indicator.

4. A second dressing later than tuber initiation is of interest to a certain extent, but too late N applications are to be avoided with respect to tuber quality and risks to the environment.

5.2   Irrigation methods of the potato in Europe

F. Martins

In this paper, potato irrigation methods across Europe are described. What are the various irrigation methods available to the potato farmer? What are the general effects of irrigation practices on crop quality, and what diseases, in particular, are promoted or avoided by irrigation? What are the cost implications of adopting a particular system? Which are the systems favoured in particular countries and regions, and what specific growing conditions characterise them? How well developed are farmer irrigation advisory systems in different countries?

Introduction

The potato can be shallow rooting where soil conditions prevent penetration by the roots. There are now many published sources showing that potatoes root to 150 cm where the soil allows; and, even, that the depth to 70% of the weight can be as much as 75 – 95 cm and with good soil conditions the potato crop draws on water to a depth of 100 cm or more.

The availability of water throughout the growing season plays a very important role - increasing quality of tubers, helping avoid certain potato diseases and some tuber defects, improving and increasing yield etc. This section covers the benefits of irrigation and describes the different irrigation methods normally used for the potato crop.

As there are substantial differences from country to country in the European Union, in terms of irrigation methods used, water requirements and the systems which exist to advise farmers when to irrigate the potato crop, some general information is presented in paragraph 3 in this subchapter, both on these issues and on the growing conditions for the potato crop in these countries. The differences referred above are very important, as they determine the demand for a given irrigation facility, its fixed cost and the amount of water used and, as a consequence, the production cost of the crop.

Effect of water on crop quality and disease susceptibility

The main effects of irrigation on tuber quality and disease are summarised in Table 1.

Table 1   *Influence of water on quality of potato tubers.*

Positive effects

Avoiding diseases
Common scab: In some years, scab control is as important a part of the irrigation regime as is growth response. In some seasons, scab control must be considered before the time of the first application of water for yield (Foley, 1987).

Reducing physiological defects
Tuber Number: This can be influenced by the water stress during the period of tuber initiation; the longer the period of stress with the soil water potential below -25kPa, the greater the reduction in tuber number per stem (MacKerron, 1987).
Cracking, Secondary growth, Chain tuberization: intermittent water supplies after tuber initiation can have almost as deleterious an effect on the crop as drought can, resulting in mis-shapes, second growth and growth cracks (MacKerron, 1987).
Jelly end rot: The incidence of this disorder is strongly influenced by seasonal conditions, being a severe problem in some years and not a problem in others. It is related to high field temperatures accompanied by drought or insufficient irrigation in the early growing period, followed by favourable temperature or sufficient water to stimulate tuber growth (Hooker, 1983).

Effects on composition
Improving dry matter content: it is influence by a wide range of factors affecting growth and development of the crop including available soil moisture (Storey & Davies, 1992).
Sugar content: Intermittent water supplies after tuber initiation can have almost as deleterious an effect on the crop as drought can, resulting in mis-shapes, second growth and growth cracks; these tubers, specially with second growth, have a higher content in reducing sugars than normal tubers (MacKerron, 1987).
Improving storage properties: Weight loss from non-irrigated potatoes were slightly higher than from irrigated ones (Ward, 1987).
Prevention of bruising during the entire process: Growing conditions which increase tuber susceptibility to cell wall fracture and external cracking (Storey & Davies, 1992).

Table 1   Influence of water on quality of potato tubers (cont.).

Negative effects

Powdery scab: is traditionally a problem of heavy soils in cool areas, that is, in soils that retain moisture at levels, which allow tuber infection. Irrigation increases tuber infections even in light soils. When irrigation is used it is especially important that only healthy seed tuber are used (Foley, 1987).

Late blight: Overhead irrigation of crops can provide favourable conditions for the development of blight in fields with a thick canopy, which also assists in the dispersion of the infection. To avoid this, the irrigation should be done during a period of the day that allows the foliage to dry, thus preventing the infection from developing. In any case, the best protection of foliage should result from the continued presence of an appropriate level of fungicide on or within the leaf/stem during irrigation (Foley, 1987).

Black leg: Irrigation water itself may carry these bacteria and cause bacterial numbers to increase; rainfall prior to lifting is also an important factor in this respect (Foley, 1987).

Rizoctonia: High moisture levels in soils, especially those poorly drained, also tend to increase severity of sclerotial formation on new tubers (Hooker, 1983).

Seed pieces decay: Saturated soils favour seed piece decay caused by bacteria and by *Pythium* (Rowe et al., 1993)

Irrigation Methods

There are basically 3 methods of irrigating potatoes or any other crop: surface, sprinkler and drip or trickle irrigation. The main characteristics of these methods are as follows:

Surface irrigation

Water is applied to a flat or uniformly sloping area, which can be limited by dikes or borders. If the area is undivided, the system is called basin irrigation; if it is divided in to narrow strips, we have border strip irrigation. If the surface is ridged in small channels, the method is called furrow irrigation (Fig. 1), which is a quite common irrigation method for potatoes. Infiltration occurs laterally and vertically through the wetted surface of the furrow; systems may be designed with a variety of shapes and spacing, with optimal furrow lengths being primarily controlled by intake rates and stream size. The advantages of furrow irrigation are uniformity of water distribution when the system is adjusted for variations in the rate of infiltration, a low energy requirement, and a relatively low equipment cost. Ridgers, drawn by conventional tractors, allow greater control to be exercised over the direction and alignment of the furrows thus created, in accordance with the topography of the land being worked. Good water efficiency can be achieved by reusing the excess run-off which accumulates at the lower end of the field. The major disadvantages of furrow irrigation are (1) the relatively high labour costs, as it takes 2 to 4 hours per hectare of labour time (except with automated systems) and (2) the difficulty in managing the system in fields with substantial variations in infiltration rate. Furrow irrigation is not appropriate to coarse textured soils, due to their high infiltration rates, and can be practised only in fields with slopes of less than 3% (Curwen, 1993; Kay, 1986). Soil is compacted by the water, which creates some difficulties at harvest, due to hard soil and large number of clods, especially when organic matter content is low.

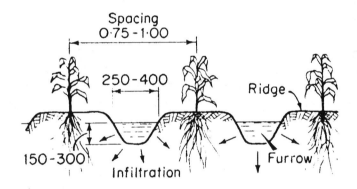

Figure 1   Furrow irrigation. A cross section perpendicular to the direction of the furrow. The distance between the middle of the furrows is in metres and the other distances in millimetres. Source: Kay (1986).

Sprinkler irrigation

Water is delivered through a pressurised pipe network to sprinklers, nozzles or jets, which spray the water into the air. This spray then falls to the ground as artificial 'rain'. The basic components of any sprinkler systems are: a water source, a pump to pressurise the water, a pipe network to distribute the water throughout the field, sprinklers to spray the water over the ground, and valves to control the flow of water. The sprinklers, when properly spaced, give a relatively uniform application of water over the irrigated area. The main type of sprinklers systems are:

Manually-Moved or Portable Sprinkler System

These systems employ a lateral pipeline with sprinklers installed at regular intervals. Traditionally, the lateral pipe was often made of aluminium, with 6, 9, or 12 metre sections, with special quick-coupling connections at each pipe joint; the same type of pipes can be produced in 6 m length plastic (PVC), or galvanised steel. Modern systems use flexile lateral pipes.

The sprinkler is installed on a pipe riser so that it may operate above the crop. The risers are connected to the lateral at intervals to correspond to the desired sprinkler spacing. The sprinkler lateral is placed in one position and operated until the desired amount of water has been applied. Then the laterals are placed in another position and irrigation repeated, and so on. Subsequently the lateral lines are then disassembled and moved to irrigate a further location. This type of sprinkler system has a low initial cost, but a high labour requirement. It can be used on most crops, but the laterals can become difficult to move in mature dense foliage. On bare 'sticky' soils, moving the lateral lines is very difficult, and extra lines ('dry' line) are used (Solomon, 1988). The cost of labour to move the pipes and sprinklers in a mature and wet potato crop is very high (up to 350 man minutes per hectare, almost 6h/ha) as well as being an unpleasant task (Bailey, 1987).

Some crop damage is also unavoidable with this system. Adding fertiliser to the water can also be considered, as there is little run-off wastage associated with this system, on all but the very heaviest of soils. Costs for in field equipment are high, typically 1700 to 2500 GBP per hectare. To reduce the amount of work required to move pipes and sprinklers from one position to the other, there are alternative variations to this method.

Solid Set

Solid set systems are similar in concept to the hand-move lateral sprinkler system, except that adequate number of laterals are placed in the field so that it is not necessary to move pipes during the season. Control valves are used to direct the water from which irrigation water is required into those laterals at any particular moment. The pipe laterals for the solid set system are moved into the field at the beginning of the season (after planting and perhaps during the first stages of cultivation), and are not removed until the end of the irrigation season (prior to harvest). The solid set system requires labour to be available at the beginning and end of the irrigation season, but minimises labour needs during the irrigation season.

Permanent Systems

A permanent system is a solid set system where the main supply lines and the sprinkler laterals are buried and left in place permanently, i. e. from season to season, this is usually done with PVC plastic pipe.

Side Roll System

The lateral line is mounted on wheels, with the pipe forming the axle (specially strengthened pipe and couplers are used). The wheel height is selected so that the axle clears the crop as it is moved. A drive unit, usually an air-cooled gasoline-powered engine located near the centre of the lateral, is used to move the system from one irrigation position to another by rotating the wheels, after first draining the axle pipe of water (Solomon, 1988).
The sprinkler system requires a large investment. The quantity of labour needed decreases from the manually-moved to the permanent system, which is particularly suited to automation and permits small amounts of water to be applied when needed. The water distribution is normally good.

Travelling Gun System (Hosereels)

This system utilises a high volume, high-pressure sprinkler (or 'gun') mounted on a carriage, with water being supplied through a flexible hose. The gun moves continuously, irrigating as it moves. The carriage may be moved through the field by a winch and cable or, more commonly, the hose is unreeled as the tractor pulls the carriage across the field. The gun used is usually a part-circle sprinkler, operating through 180 to 210 degrees of the circle for optimal uniformity of irrigation while allowing the carriage to move ahead on dry ground. These systems can be used on most crops, though due to the large droplets and high instantaneous application rates produced, they are best suited to coarse soils with potentially high infiltration rates and to crops providing good ground cover (Solomon, 1988).

Adequate pressure at the gun nozzle (usually 4.5 -5.5 bar) is essential to atomise water correctly. In a raingun with a 22 mm nozzle, operated at 8 bar, 78% of the droplets are in the range of 1-2.2 mm and the remaining droplets are bigger. Using a smaller than usual nozzle (18 mm) and a reduced application rate, potato crops can be irrigated satisfactorily on many soil types prior to full crop cover development. Such 'rainguns' are very sensitive to wind conditions, but booms operate well even in moderate winds. To enable rainguns to operate in light to moderate winds, a lane spacing of 70% of the potential spread width in still conditions is recommended (Bailey, 1987).

In most cases the coefficient of uniformity (CoU) is only 75%, but it can be as low as 44%. The use of variable trajectory rainguns, properly operated, can achieve a CoU of 90% both day and night, and in a range of wind speeds.

The most common fault limiting the efficiency of their use is incorrect, low pressure at the nozzle, which should be checked regularly. Spacing the passes of the raingun at 72m requires 4 – 5.5 bar at the nozzle; however, the CoU can be improved by operating at 60m spacing.

To reduce the droplet size and increase the uniformity of application of water, a boom can replace the raingun. The boom is a cable-supported framework of pipes on which sprinklers are mounted. 98% of the droplets produced by boom-mounted sprinklers will be in the range of 1-2.2 mm, thereby limiting damage to the soil structure. The use of boom irrigators has some advantages. Uniformity is very good, a CoU of 90% is consistently achievable, and they only require low pressure to operate effectively. The disadvantages of using booms include their cost and a very high, instantaneous application rate.

The capital cost of this equipment is about 15000 GBP per hosereel or 23000 GBP per hosereel with boom. With an application rate of 25 mm over 7 days, each machine can be expected to cover 20 hectares. While these machines are used in 95% of to potato irrigation in the UK., in dry areas, such as Portugal, 25 mm in seven days can be insufficient, and higher capital costs per hectare have to borne.

Centre Pivot and Linear Move Systems

The centre pivot system consists of a single sprinkler lateral supported by a series of towers (Fig. 2).

The towers are self-propelled so that the lateral rotates around a pivot point in the centre of the irrigated area. The time for the system to revolve through one complete circle can range from a half a day to many days. The longer the lateral, the faster the end of the lateral travels and the larger the area irrigated by the end section.

Thus, to deliver an even amount of water, the greater the distance from the centre, the greater must be the application rate. The high application rate at the outer end of the system may cause runoff on some soils. Since the centre pivot irrigates a circle, it leaves the corners of the field unirrigated (unless special attachments are used). Centre pivots are capable of irrigating most field crops (Storey & Davies, 1992), and produce small droplets, averaging 1.3 mm according to Bailey (1987).

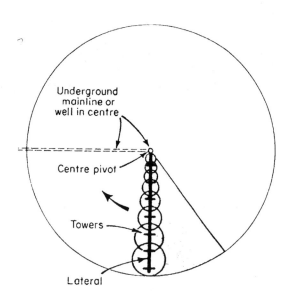

Figure 2   Centre Pivot (Source: Kay, 1986).

Linear move systems are similar to centre pivot systems in construction, except that neither end of the lateral pipeline is fixed. The whole line moves down the field in a direction perpendicular to the lateral. Water delivery to the continuously moving lateral is by flexible hose or open ditch pickup. The system is designed to irrigate rectangular fields free of tall obstructions. Both the centre pivot and the linear move Systems are capable of very high efficiency in water application. They require high capital investments, but have low irrigation labour requirements. The distinct advantage of these irrigators is that water can be applied 'little and often' with the minimum of labour and inconvenience (Solomon, 1988).

Normally, the higher the development of a country the wider is the range of technical choice. An important factor is the energy cost of pumping the water. Farmers normally choose systems with lower pressure requirements, with lower labour requirements, and try to irrigate fields where electricity is available.

Acording to Kay (1988), when choosing a sprinkler system, many factors must be considered:

- If the land slope is less than 5%, any type of system can be used; with slopes greater than 15% only conventional portable, permanent and semi-permanent systems will be suitable.
- All the systems described above are easily adapted to regularly shaped fields such as squares or rectangles, conventional, mobile rainguns and small side moving systems can be adapted to irregularly shaped fields; centre-pivots only irrigate circular areas and as much as 20-25% of a square field will remain unirrigated unless a special corner irrigation device is available.
- The sprinkler systems can be easily adapted to the soil infiltration rate; if the soil surface structure is easily damaged under irrigation (capping) then mobile rain guns should not be used.
- The choice of sprinkler irrigation methods is based on several criteria, some of which are summarised in Table 2.

LEPA Systems: Low Energy Precision Application (LEPA) systems are similar to linear move irrigation systems, but are different enough to deserve separate mention. The lateral lines are equipped with drop tubes and very low

Table 2   Comparison between sprinkler irrigation methods.

System Type	Capital Cost	Energy use	Labour	Main-tenance	Running costs	Irrigation efficiency
Manually-Moved or Portable	L	L	H	L	L	65-75
Solid Set	H	L	M	L	M	70-80
Permanent	H	L	L	L	L	70-80
Side Roll	M	L	H	L	L	65-75
Travelling Gun	M	H	M	H	H	60-70
Centre Pivot (C.P.)	M	L	L	M	L	75-90
C. P. With corner system	M	M	L	H	M	75-90
Linear move (ditch fed)	M	M	L	H	M	75-90
Linear move (hose fed)	H	M	L	H	H	75-90

H – high, M – medium, L – Low. Source: Based on Solomon (1988).

pressure orifice emission devices which discharge water just above the ground surface into the furrows. This distribution system is often combined with micro-basin land preparation for improved runoff control (and to retain any rainfall, which might occur during the season). High efficiency irrigation is possible, but these systems require either very high soil intake rates or adequate surface storage in the furrow micro-basins to prevent runoff or non-uniformity along the furrow; the irrigation efficiency is very high, 80-95% (Solomon, 1988).

Drip or Trickle irrigation

Trickle irrigation is the slow, frequent application of water to the soil though emitters placed along a water delivery line. The term trickle irrigation is general, and includes several more specific methods. Drip irrigation applies the water through small emitters to the soil surface, usually at or near the plant to be irrigated. Subsurface irrigation is the application of water below the soil surface. Emitter discharge rates for drip and subsurface irrigation are generally less than 2 litres per hour (Solomon, 1988). It is used in potato production in a few areas where water is in short supply.

Pipes for drip irrigation are made of polyethylene (PE), with the emitters incorporated in the pipe, especially when used to irrigate row crops. In some cases, the wall of the pipe is very thin, so that, when empty, pipes remain flat and resemble a tape. This kind of tape is typically used for potatoes; it is placed at the top of the ridge and buried 2 to 3 cm under the surface. Placing and retrieval of the tape can be done mechanically.

Another option is to locate a single line in the middle of a two-row bed. However, it is difficult to ensure that the water will move laterally. If the soil is too wet, the water tends to soak in vertically; if too dry, the water also tends to leave the beds unirrigated.

Use of trickle irrigation may save up to 30% of the total water applied, but only if it is planned and timed appropriately, otherwise it can use more water, particularly if irrigation is automated.

There is a choice of weight between tape and tube, with the heavier material costing more but, potentially, lasting longer. In Table 3 the advantages and disadvantages of drip irrigation are presented.

Table 3   Advantages and disadvantages of drip irrigation.

Advantages	Disadvantages
Water use efficiency, water saving	High installation costs
Excellent uniformity of distribution	Needs more control for effective management
Better timing – control, timing, and quantity	May cause restricted root development
Improved plant response and, possibly, yield	May cause poor seed germination
Improved application of chemicals / fertilisers	Risk of over-irrigation
Reduced energy requirements	High maintenance costs – clogging – if re-used.
Can be automated – reduced labour (peak at laying and lifting)	Needs appropriate planing and timing
Accuracy in volume and frequency of application of water	It is necessary to ensure that leaching does not occur
Flexible with regard to soil type and topography	

Based on Curwen, (1993).

Climatic conditions for the potato crop in Europe

Conditions for growing potatoes vary throughout Europe, with diverse water requirements and the use of different irrigation methods characterising specific zones and countries. General information about the yield and the growing conditions of the potato is presented in Table 4.

Country-specific problem(s)

In Portugal the biggest share of the production, 49,1% (Rolo, 1995), belongs to the northern regions near the coast. These two regions have a strong Atlantic influence, with a cooler damper summer, than in the rest of the country, especially in the coastal area. The irrigation needs are small, but it is necessary to irrigate, especially if the crop is planted late or if a main variety is used. As the Atlantic influence is absent and the period of frosts consequently longer in the Trás-os-Montes region (responsible for 25.6% of national production) (Rolo, 1995), it is necessary to plant late, the crop grows in much higher temperature and the need for irrigation is increased. In both Western coastal and North interior regions, if the spring is wetter than normal, there are very strong attacks of late blight, with severe losses in the crop.

In most years, the southern and eastern districts of England receive less summer rainfall and exhibit higher evapotranspiration rates than the rest of the United Kingdom. Consequently, these districts are more prone to drought, and account for approximately 80 % of the total UK irrigation use.

In Belgium Irrigation is poorly developed all over the country. It only represents a small proportion (no more than 5%) of the total agricultural area. Irrigation, when present, is undertaken on sandy soils in the north of Belgium or, more recently, in the loamy soils in the centre of the country, primarily for irrigation of vegetable crops, and secondarily for potatoes. In several parts of the country, irrigation is not used, essentially due to the small size of the fields and to the high-cost access to groundwater tables which are the most important water sources in Belgium.

Yield losses due to both drought and rain are common in Finland. The main cultivation areas are located in the western part of Finland where early summer droughts often cause serious yield losses in the fine sand and silt

Table 4   Country information

Country	Yield (t.ha⁻¹)			Growing conditions							
	Natio-nal Avg	Well Irriga-ted	Not Irriga-ted	Plant Time	Harv Time	Avg. Max (°C)	Avg (°C)	Avg. Min (°C)	Avg. Trans (mm)	Rain G.S. (mm)	SAV (v/v%)
Portugal (W.Coast)	14	30-60	12-18	Jan Feb	Apr Jul	13/22	10/18	6/15	3-5	500	15 25
Portugal (Inland)	14	30-50	9-12	Mar May	Jul Set	15/29	10/22	5/14	5-7	350	10 20
United Kingdom (Eastern England)	42	40-60	6-50	Feb Apr	May Nov	7/22	4/16	0/10	3-1	300	10 23
United Kingdom (elsewhere)	42	40-60	15-60	Feb Apr	May Nov	7/22	4/16	0/10	2.5-3.1	300 500	10 23
Belgium	45	45-75	30-65	Mar May	Jul Oct	8/22	5/17	1/12	2.5-3.0	350	
Finland	20	25-50	10-35	May	Jul	15/22	2/17	-5/10	1-3.5	250 350	15-28
The Netherlands	47	50-60	30-60	Apr	Aug Oct	3/21	2/17	-1/13	3-4	300 400	8 25
Poland (C. Part)	20.3	30-50	8-30	Apr May	Jun-Nov				2.0 3.7	300 400	10 24
Spain (Galicia)			20-30	Mar May	Jul Sep				2-3	580	
Spain (Central)									3-4	220	
Spain (Andalucia)									3-5	240	
Italy (North)	26	32	18	Mar May	Jul Sep						
Italy (South)											

Notes: Avg. Max (°C); Avg (°C) Avg. Min (°C): temperature Avg. Transp (mm), transpiration, and Rain G.S. (mm), during the growing station. SAV (v/v%): soil available water.

soils. The irrigation need for potatoes is on average 100 mm during the growing season. Irrigation is used in approximately 15 % of the total potato cultivation area. Furthermore, irrigation is also used to prevent damage caused by frost.

The most common irrigation method in each country is presented in Table 5.

Table 5   Most common irrgation method in each country.

Country	Furrow	Rainguns	Sprinkler	Centre Pivot	Trickle
Portugal (W.Coast)	++++	+	+++	0	0
Portugal (Inland)	++++	+	++	+	0
United Kingdom (Eastern England)	0	++++	+	+	+
United Kingdom (elsewhere)	0	++++	+	+	+
Belgium	0	++++	+	+	+
Finland	0	+++	+++	0	0
The Netherlands	0	++++	+	+	+
Poland (C. Part)	++	+++	+++	+	+
Spain	++	+	+++	+	+
Italy	+	+++	++	0	0

key: very common (++++) to less common (+); 0 = rare/absent

Estimation of irrigation needs

Currently, no estimates of evapotranspiration are available for Portuguese and Polish crops. In the UK there is a comprehensive network of meteorological stations throughout the country, including a large number in the area where irrigation is most commonly used and evapotranspiration estimates based on data from these stations are widely available. There is a similar network of meteorological stations throughout Belgium, providing on request data from which daily evapotranspiration estimates may be calculated, but limited use is made of it, as irrigation is still a relative rarity. The Finnish Meteorological Institute also has a specific telephone weather service for farmers. However, estimates of evapotranspiration of the crops based on data from the network of meteorological stations of Finland are not available; maximum required irrigation is approximately 100 to 150 mm. In the Netherlands the meteorological institute includes in its weather forecast for farmers each day the so called reference evaporation (reference to a low cut grass field); Farmers can use these data (broadcast on radio or available via a commercial computer service) when they apply the 'balance method' for estimating irrigation need. Some growers use tensiometers. While there is no special advisory service for irrigation practice, the National Advisory Service does promote the use of the 'balance sheet' method. No more than 10-20% of farmers are using either the balance sheet method or tensiometers; however there is a tendency to increase the use of both, especially in regions where water is scarce.

Farmer Advice System for Potato Irrigation

In Portugal there is no advisory service for irrigation. In the coastal region, crop water requirements are almost entirely satisfied by rainfall. In other areas, farmers irrigate once a week or every ten days, if water is available. In areas where water is scarce and several farmers share access, the irrigation of the crops is undertaken whenever water is available.
In the UK there are several organisations offering irrigation advice to growers, based upon Penman or Penman-Monteith evapotranspiration estimates. Such services are run commercially and farmers are charged a fee. The ADAS Irriguide is the most widely used, with 300 farmers subscribing to the service. Also some organisations offer chargeable advice based on soil water measurements, usually via neutron or capacitance probe. Despite the wide availability of irrigation scheduling services in the UK, probably no more than 15% of farmers take advantage of them. Most prefer to irrigate according to a fixed time schedule, incorporating delays during periods of high rainfall.
In Finland in 1996, the Potato Cultivation Institute published a report on the irrigation of the potato crop. The document includes demand for water at different crop stages; estimates of irrigation need on the basis of precipitation

shortfall, soil water capacity or soil moisture content. There is no advisory service for irrigation operated either by telephone or computer.

In Poland no Advisory Service for Irrigation exists. Most of the crop's demand for water is met by rainfall. In the small areas where potatoes are grown exclusively for 'French Fries', farmers irrigate once a week or every ten days, according to water availability.

Conclusions

Irrigation is a very important practice during the growth of the potato crop, as it plays a key role in helping to avoid diseases and contributing to a better quality product, in terms of dry matter content, shape and size of the tubers harvested. Also it has a significant influence on the yield of the crop, especially in the southern part of Europe, where Spring and Summer are hot and dry and, if the crop is not irrigated, the yield tends to be very low, and production costs very high.

Irrigation can be effected by several methods, chosen on the basis of several factors, mainly land topography, soil infiltration rate, field shape, labour requirements, capital cost, energy use, maintenance and irrigation efficiency.

Recommendations

- Surface methods, specially furrow irrigation, is cheap, and the furrows can be easily made by the farmer with a tractor-drawn ridger.
- Sprinkler systems need substantial investment; running costs are expensive and it is almost always necessary to have an electricity supply. Water distribution is good and the equipment, mainly pipes and sprinklers, can be increased over time, avoiding large initial simultaneous expenditures.
- Hosereels with a rain gun or a boom, are very effective irrigation systems when the total amount of water applied is low. They are sensitive to wind conditions, particularly with raingun and the running costs are high due to the high-pressure needed to produce small droplets.
- Centre pivots need more investment and larger areas to irrigate, as approaching 40 ha can be irrigated even by one of the smaller units. Electrical power supply is indispensable.

- Drip or trickle irrigation is very useful for regions with habitual water shortages; its main inconvenience is the installation cost, and the necessity of removing the tape before harvesting.

5.3   Role of simulation and other modelling approaches in decision making

T. Karvonen, D.K.L. MacKerron & J. Kleemola

What is meant by a simulation model? An expert system? Or a decision support system? What are the new aspects models can bring? What kind of decisions can computers help with? What kind of input data is needed? Who collects it? Where does the weather data come from? Who makes the simulation runs? Is it possible to accommodate environmental aspects of growing crops and still obtain high yields? What is the role of a farmer in a computed assisted decision? Is the computer making the decision?

Introduction

Mathematical models and expert systems have been developed and used by scientists for at least three decades. An important reason for developing models is that they should be able to help a farmer make decisions. However, an ordinary farmer or a person not familiar with models is probably very uncertain, confused, and even afraid of the terminology surrounding models and quite unconvinced of their applicability to farming. The farmer influences the performance of his fields by adjusting management practices, in fact, manipulating the controllable components of the production system to attain his objectives as fully as possible. Models can be developed that will allow the farmer to explore his options before actually committing resources. Various kinds of models, expert systems and decision support systems (DSS) will be explained here and indications given of their suitability to aid environmentally friendly potato production. The main goal of this chapter is to give an overview of the role of simulation models and expert systems in crop production.

Strictly speaking, a simple table with rows and columns labelled by soil type and previous crop, and each cell showing the required amount of nitrogen per hectare) is a support for the farmer deciding how much fertilizer to give his crop. However, when we think of decision support systems we think of the use of computers that have been programmed in some way to take in information from several sources, weigh up possibly conflicting requirements, and provide an answer. Whether the necessary information will be readily available and just how correct will be the answer, are matters that we will explore in this chapter.

It is necessary to distinguish between the different types of models available and to clarify the basic terminology. Then we discuss the use of models and expert systems as aids for long-term (*strategic*) and short-term (*tactical*) decisions. Strategic decisions on agro-ecosystems are based on the expected long-term performance of these systems. Tactical decisions are those taken in response to the actual state of the system, the farmer's crop, and its environment at any particular moment. The time horizon of tactical decisions is, therefore, within-season even, in some cases, only a few days.

Modelling can quantify how much yield, for example, will change if an action is taken, or not taken. The use of modelling in agricultural DSS is still relatively new and, although it offers promise for the future, several questions need to be answered. One important topic is to consider who is foreseen to run the models: the farmer himself, or an advisor, for this will affect what is an acceptable system. Other questions include, who collects the input data and where are the decisions made? Another consideration is that as farming involves several forms of uncertainty, decisions should be accompanied by an analysis of risk. We will argue that a DSS must never be allowed to 'make' the decisions. It will be important to remember that such systems are designed for support. We discuss these questions and the likely future developments of DSS such as the combination of the environmental aspects, profit, and crop quality in modelling and decision making. Another possible development is to combine on-farm, real-time, field observations with simulation modelling and expert systems to suggest the most appropriate management practices to achieve a desired effect.

Classification of mathematical models and other modelling approaches

Mathematical models and related approaches can be classified into several groups: *statistical regression* models, *dynamic simulation* models, *optimisation* models, *expert systems,* and *decision support systems (DSS)*. Further, a distinction is necessary between *research* models and the simpler *production versions* of them that can be used in DSS. Statistical regression methods describe relations between a few explanatory variables e.g. sums of intercepted solar radiation and rainfall during the growing season, and the variable of interest, e.g. yield. The outstanding characteristic of regression models is that the relations they describe are only necessarily contained in the data from which they were derived. There can be no confidence that they will be applicable across other data sets, with different input values. Such models are often used in policy making but they are not suited to tactical or operational decisions and so we will not consider them further.

Dynamic simulation models, often called deterministic or mechanistic, have two basic features that are important from the decision-making point of view. First, they attempt to explain the causal relations between environmental variables, growth functions, and crop yield and second, as 'dynamic' implies, variation in time is a dominant feature of the data. Dynamic models compute state variables (i.e. 'what is the position now') such as biomass, soil water content, soil nitrogen content, etc. These models need input data, driving variables, that are measured as functions of time. The most important driving variables for growth are rainfall, air temperature, solar radiation, and potential evapotranspiration. The mathematical equations used to calculate the rate of change in each state variable include values called parameters. In some cases, parameter values will be cultivar dependent and parameters for available nitrogen in the soil are closely site-specific. Therefore, it is often necessary to 'calibrate' a model for a given location. The calibration has to be done by an expert and not by the farmer himself. In addition, if at all possible, the model should be validated (verified) against an independent data set. Simulation models can be used both in strategic and in tactical decision making. The strength of the simulation models is that they allow analysis of 'what if' questions, i.e., what would be the result if a specific action were taken at this moment, postponed, or not taken at all? In some situations they also have a role in explaining or understanding a past event or an outcome, and so can help with similar decisions in the future.

Optimisation models go a stage beyond the simulation models. Their purpose is to maximise, or minimise a target function by changing certain decision variables. To explain: a target function is usually something with an economic or regulatory value, e.g. yield, and a decision variable is a quantity that the farmer can influence, e.g. amount of irrigation or fertiliser to be applied. Usually optimisation is designed to satisfy some constraints, possibly conflicting, e.g. water for irrigation is limited and several crops require it; or nitrogen is needed for growth but there is a maximum permitted level of residue. Optimisation models are most useful helping strategic decisions. They can also be used for tactical decisions but only if the necessary data is available. For example, data on average rainfall is appropriate to guide strategic decisions on irrigation but inappropriate when considering action at any particular time.

Unlike simulation models, expert systems are intended to work in the way that a human expert might think. They recognise two kinds of knowledge - domain and strategic. Domain knowledge is what most people think of as knowledge and strategic knowledge is knowing how to use the first kind of knowledge. This may seem complex, and it is, but so are human thought processes. An 'expert' will have a set of rules so that in one set of circumstances he uses one set of relations and in another set of circumstances he uses others. So, expert systems formulate existing knowledge into simple 'if...then' rules. These systems can then also use simulation and optimisation models of crop growth. A well-designed expert system uses all the information previously available together with current data to recommend an action.

Several expert systems have been developed during the last decade for a range of applications in both strategic and tactical decision making.

Decision support systems are very closely related to expert systems but use a much larger range of methods. DSS combine data, information and knowledge that are available in databases, mathematical models, and information provided by the users themselves. Decision support systems usually also include methods to handle uncertainty and various forms of risk analysis. DSS, too, can help in making both strategic and tactical decisions.

The main, important difference between simulation models and DSS is that the latter suggests an action whereas the former calculates what happens if a certain action is taken.

Use of models for strategic decisions

Strategic decisions on crop production are based on long-term performance in production and likely weather. The time horizon for such decisions is of the order of several years. Examples of strategic decisions in potato growing are the purchase of new machinery, or irrigation and drainage systems. The purpose of using models or DSS in making such decisions is to provide an assessment of risk to attach to a cost-benefit analysis:

Example 1: In Northern Europe the growing season is very short and drainage systems are needed to ensure that the fields are trafficable soon after snow-melt or winter rains. The cost of installing drainage is high compared to yearly profit from the crop and is increased by closer drain spacing. Yet planting is later and yields decline with wider spacings. The strategic decision to purchase or not requires information on whether there is a drain spacing that would allow profit to exceed costs over a period and, if so, what drain spacing would maximise the profit over a period of years. The task is to find a drain spacing that on the average gives the best margin between income and amortised costs. Modelling offers the means to derive a solution providing (i) the influence of drain spacing on soil water can be quantified, (ii) a model is available to simulate crop growth influenced by water, (iii) historical weather data from a period of, say, 20 - 30 years is available, (iv) drainage costs and crop prices are known and interest rates can be estimated. Then the simulation model can be run for a range of possible drain spacings and for all the years in the historical weather data set. This allows calculation of average income and average amortised costs for several drain spacings and it is easy to calculate which spacing gives the best profit. More importantly, it also gives information on the possible medium-term outcomes in years as extreme as any in the data set. So, risks can be assessed.

A comparable strategic decision on manuring might be to assess the consequences of increased use of organic manures.

Use of models for tactical decisions

Tactical decisions are taken in response to the current state of the field or crop and the perceived future. The time horizon of these decisions ranges from a few hours to some weeks. The degree of success in potato production during each growing season depends largely on the quality of the tactical decisions. The theoretical basis of applying models to tactical decisions has already been explored (van Keulen & de Vries, 1993).

When discussing the application of models and DSS in decision making it is useful to define two different levels of production, potential and constrained. Modellers may debate what is potential yield but here we will mean simply that yield that is achievable by a crop given actual planting and harvest dates and the current season's levels of temperature and radiation. Calculation of potential yield assumes that all constraints of water and nutrient supply, pests and disease are eliminated. Generally, crops achieve less than their calculated potential - but not always. During the growing season, the crop moves from crisis to crisis and it is the job of the crop manager, the farmer, to minimise the effects of these crises. They are avoided where the farmer manages to remove constraints on growth and development. Apart from pests and disease, the most common constraints are caused by imperfect supply of either water or nutrients and there are several models available that simulate the effects of each of these individually. But, as already stated, real crops will suffer a sequence of constraints. The different models have common needs for input data - weather - and unique needs, peculiar to the constraint being considered. The most common constraint to be simulated is water supply and there are a number of models for this purpose. These models need to calculate growth under water-constraint and also under potential conditions. Thus they can indicate the benefit from removing the constraint. Crop growth must be simulated and a soil water balance maintained. The principal data required are: planting date, temperatures and solar radiation for the calculation of potential growth and rainfall, potential evapotranspiration rate, and soil type and drainage for the water balance and to estimate severity of the constraint.

Example 2: A farmer wants to schedule irrigation to derive maximum benefit from a previous investment in irrigation equipment (a strategic decision). Soil characteristics including depth of topsoil, soil texture or available water capacity of topsoil and subsoil are necessary and these, together with cultivar and planting date need be given only once. In contrast, the soil water balance is dynamic and requires daily data, at least. The soil water balance is commonly summarised as a soil moisture deficit (SMD) that quantifies how much of the soil's reserves of water have been evaporated from the crop and soil and have not been replenished. Soil moisture deficit is a good example of a state variable. It defines the conditions now. It is commonly used as a flag to indicate when the crop should be irrigated. Here we come to an important distinction between a model and a DSS. Generally, a simulation model will need to be given a value for the threshold soil moisture deficit at which it should call for irrigation. In a DSS, on the other hand, the system would also consider the cost of that irrigation versus the benefit to the crop and the implications for water total water abstraction over the season. A DSS would also consider the forecast weather and the practicalities of cycling irrigation around the farm. Most importantly, it would explain the basis of its recommendations. Only the farmer knows whether he believes the weather forecast.

It is important to note that although such models are in regular use throughout the world, farmers do not normally run them themselves. Usually, farmers get the answers from service providers. This illustrates the issues of who runs the simulations, where the data comes from, and who makes the decisions.

Decisions beyond the scope of simple models

Managing the possible constraints of nitrogen supply is a more complex task. Where there is a significant quantity of organic matter in the soil, it will provide a variable amount of mineralized nitrogen to the plant. The same is true of organic manures, and mineral fertilizers are incorporated into the microbial fraction of the soil. Yet, even if the available nitrogen can be calculated, it will not be accessible unless the water supply is within certain limits that also need to be calculated. What further complicates the task of optimising the

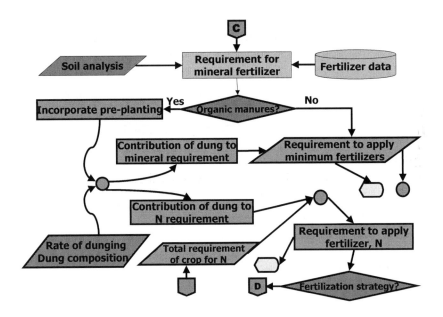

Figure 1.   A segment from a decision tree reflecting the structure of decisions on the management of nitrogen application to potato.

application of nitrogen fertilizers to a crop such as potato are the multiple objectives of balancing crop yield and crop quality while minimising environmental impact - principally by minimising residues.

The complexity of this task is such that the simpler forms of simulation models cannot be expected to provide realistic solutions. The advantages and disadvantages of alternative decisions need to be considered explicitly in quantitative terms. Recommendations are required on both the level and timing of application of fertilisers. Although there are models available that treat some of the issues involved, (Fig. 1) there is an urgent need for these to be developed into a refined system for recommendation as environmental effects are gaining in importance while profit is essential.

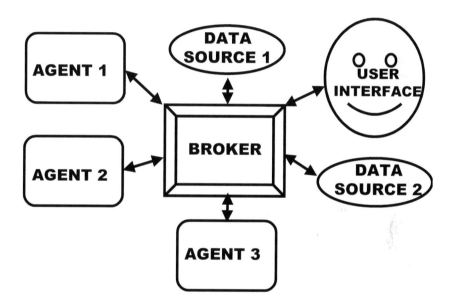

Figure 2.   A schematic representation of a Broker-agent architecture.

This is the sort of problem that is suited to solution by an 'expert system'. One possible design would be to use the expert system as a broker between the user and agents, which are sources of data or are other models. Such a so-called 'broker-agent' architecture (Fig. 2) could integrate mathematical models within a heuristic framework, provide the means to interpret simulation results and create explanations, and present information tailored to the user's needs. Tasks set by the user would be interpreted by the expert, broken down into subtasks, and allocated to appropriate agents. Significantly the farmer does not need to know the models (Fig.3).

Figure 3.   The farmer does not need to know the models.

The problem of providing a refined recommendation system is aggravated by the lack of fast, simple methods to monitor the nutrient status of the crop and soil during growth, although techniques are available that will allow spot checks. (See earlier chapters in this manual). Also, the supply of nutrients represents a commitment for several weeks ahead and so an accurate forecast of the weather for that period would be necessary for an accurate calculation of the crop's demand for them. Some reports suggest that improved methods for fertiliser recommendation systems awaits the development of better measurement techniques (van Keulen & de Vries, 1993). An alternative approach is that other less deterministic decision support can be developed (Marshall, Crawford & McNicol, 1994, 1995).

Information techniques that are probabilistic rather than deterministic will probably best be used to estimate quantities such as leaching to ground waters and rates of denitrification and immobilisation.

It has been argued that. if reliable and easy systems for crop monitoring were to be developed it might be possible to base operational practice entirely on them and disregard models. Until then, it will be important to develop the information technology at our disposal to give the best recommendations (Yiqun Gu et al., 1996).

Who runs the simulations? Who makes the decisions?

Computers are becoming ever more widely used in farming. So, it is estimated that, in the UK, 60% of farms of over 20 hectares will have a computer in 1999, and 75% of arable farms of over 150 hectares. Further, it is estimated that 60% of these computer owners use them at least daily for business (Agridata, 1999). Yet farmers make very selective use of that equipment. They are familiar with its use for record keeping and for financial management but its use in decision support is not common; even for irrigation scheduling as described earlier. So, where a computerised system is available, several important questions remain to be answered. Anyone can buy carpenter's tools, but very few can make good furniture. It is the same with models, and farmers know it. A simulation model is only a tool and a tool must have an experienced user working regularly with it who knows its strengths and weaknesses. The existence of a model is a necessary condition but not a sufficient one for a computerised support system.

The successful use models demands calibration of the parameters of the model. Who is to do this to give confidence in it? Who runs the simulations during the growing season? Where does the data come from? Who makes the final decision? While the successful use of an expert system requires that the farmer should feel comfortable with it, the problem of data acquisition does not go away.

Technically a farmer could make 'production runs' of a model during the growing season but where does he get the input data? Decisions on irrigation require weather data. Some farmers may own and run a small, automated weather station but otherwise weather data is expensive when bought by a single user. This and the time taken to acquire the data for oneself, are the reasons why most farmers will buy such advice from a service provider. On the other hand, this route does not relieve the farmer of all effort and responsibility. There are still crop and site-specific data that he will have to provide. Summer rainfall is frequently localised so it must be measured on each field. Where the crop or soil are monitored for nitrogen levels, the farmer must make the measurements. In each of these cases there has to be a two-way exchange between the customer, the farmer, and the service provider. The farmer providing selected data, the service provider providing recommendations and, ideally, supporting these with other information.

Example 3: In example 2 earlier, we considered the use of a model to schedule irrigation. In practice, the farmer measures rainfall in his fields and sends that and information on crop cover, tillage operations, and any irrigation to his service provider. That organisation is responsible for running the model that is used and for acquiring the necessary weather data other than rainfall. The advisor derives estimated soil moisture deficits for each field and suggested requirements for irrigation in the next week, based on soil moisture deficit, criteria on the deficits appropriate to the development of the crop, and a weather forecast. The suggested irrigation scheduling can be sent to the farmer. Typically this is done weekly.

A system such as this would be well suited to producing recommendations on the application of nitrogen fertilizer if a suitable decision support package were available to the crop consultants. It would even be amenable to providing answers of differing precision dependent upon the level of crop monitoring that was possible and done on each particular crop.

Recommendations may not be adopted. Just as the recommendations of the DSS are based on data that the farmer could not get, so the farmer has other information that the DSS cannot handle. The decision must belong with the farmer but, it is also important that results of any decision taken are part of the next input to the DSS.

Future developments

Already the bigger, more advanced farms have their own mini-weather stations installed; most have computers available. It is likely that these capabilities will become more widespread. In future, a farmer with such equipment may want to run his own support package. Unless extremely robust packages have been developed, the models within the system will probably require calibration by an expert and the results of the models should always be checked against the real data. The incorporation of that check data would offer great scope for improvement, at least in risk analysis, for example by providing expected ranges of outcome as well as an average or 'typical' outcome. Managing such a system is, again, likely to remain in the hands of experts.

Conclusions

- Experience has shown that most successful use of mathematical models in tactical decision found where field observations (made by the farmer) are combined with computerised systems. Crop irrigation scheduling and pest and disease control are probably the areas where models and decision support systems have been most useful to date.

- Throughout each of the countries of Europe farming is practised at a range of technical expertise. A decision support system for the management of the water and nitrogen supply to potatoes in Europe must be capable of being operated in advisory systems appropriate to each of those technical levels. But in all cases they should incorporate a requirement for data that is obtained by monitoring the crops in question.
- Models have not yet significantly improved the quality of the fertiliser recommendations. Measurement techniques and quality of long-term weather forecasts have to be improved before conventional modelling can offer more reliable recommendations.
- DSS should include estimates at least of the risk that the weather differs from expectation. For more sophisticated users that could be extended to include the possibilities that other inputs are wrong or inappropriate.
- DSS should provide estimates of the consequences of differing actions - at least of Action versus Inaction. Further, DSS should offer the chance to ask, 'What if ....?' For example, what if less is given? Or more?
- If a DSS is to be successful it must be accepted by the intended users and to that end it must address the needs and opinions of those users. The only reliable way of ensuring that is achieved will be to consult panels of users, reiteratively, as the system is being designed.
- The design of a DSS should pay particular attention to requirements for data, possible sources of that data, and the cost of its acquisition
- Information required should include limits to the amount of irrigation that is available, prices of inputs, the value of the output from the crop, and penalties associated with residues.
- Decision support should not be confused with the decision itself.

6   Future developments

6.1   Exploring trends in current practice, what are the lessons and risks?

P.A. Johnson & M. Colauzzi

Potatoes are grown on 1.1% of Europe's agricultural land. A large proportion of the total irrigation capacity of some countries is used on the potato crop. Some of the N fertilization is of an 'insurance' nature particularly where organic manures are applied. Have European potato growers taken up the irrigation scheduling services and N prediction systems which have been developed in the last decade? How much notice do they take of the nitrogen supplied by organic manures? Is there a need for more promotion of these systems as water supplies come under pressure from other users and nitrate is blamed for deteriorating water quality?

Introduction

Earlier chapters in this handbook have discussed at length many of the scientific and technical advances which would enable potato growers to use both nitrogen fertiliser and water both more efficiently and in a more environmentally friendly way. This chapter looks specifically at the attitudes of potato growers to the use of water and nitrogen and the pressures they face as water supplies become scarcer and there is a need to reduce the amount of nitrogen entering the aquatic environment. We look at examples of attitudes and use of technology using examples from Italy and England specifically to show examples from Mediterranean and maritime countries.

Water

Water supplies in many areas of Europe are under pressure due to increasing consumption by domestic, industrial and farm users. The pressures have increased because of housing development, and the need of farmers to irrigate in order to achieve the ever more stringent quality parameter required by buyers. With increasing pressures on farm incomes the production of high yields of high quality tubers targeted to particular markets is required. It has been estimated that water is the most limiting factor in potato production even during the short growing season experienced in Finland. It has been clearly shown that irrigation will increase yields and quality. Thus interest in irrigation is expected to increase in the future both from the point of view of crop yield and of quality.

Abstraction of water for spray irrigation of agricultural crops in some areas of south east England is now around 23% of total abstraction. This has contributed to the low flow in some rivers particularly during summer. It has been suggested that demand for water for irrigation could increase by a further 50% before 2021 and that climate change could add a further 27% in demand. In Italy water use on the potato crop appears to be reducing (Fig. 1), which is unexpected. Available information shows that there is a tendency to reduce the seasonal applied amount of irrigation water in productive areas located in the north and middle of Italy.

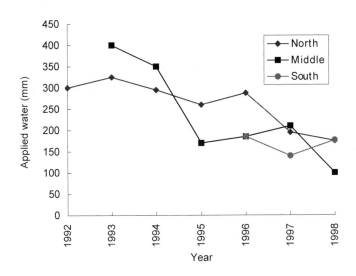

Figure 1   *Water applied to the potato crop in Italy in the last seven years. Each line represents*
              mean values 3-4 productive areas. (Source 'L'informatore agrario').

The reduction in the use of irrigation water to the potato crop in Italy is connected to the improved efficiency of new irrigation systems. In the north there is the tendency to reduce the pressure in the pipes (this implies permanent systems and smaller guns) and to use more systems based on booms that distribute the water more accurately. In the south, considering the part of the year when potatoes are grown irrigation is not always needed, especially on clay soils and in wet springs. In the typical southern areas the potato crop it is grown in rotation with various valuable vegetables. This implies a high level of specialisation of the farmers and a particular ability in the use of low-pressure irrigation systems (drip irrigation), which are sometimes used in combination with plastic mulches. In the England the area of potatoes irrigated doubled to 62110 ha between 1982 and 1995 (MAFF, 1997) with a corresponding increase in water use from 17,050,000 m3 to 88,806,000 m3 (54% of all water used for irrigation). Some 40% of the water is abstracted from rivers, with only a small proportion during winter, and 33% from boreholes.

New licences for groundwater extraction in many areas are unobtainable, only winter abstraction from rivers is allowed subject to adequate river flows. Indeed many aquifers in the south east of England have been at an all time low for some years. The problem is compounded by the fact the majority of water abstracted for irrigation occurs in a 2 to 3 month period in the summer when evaporation rates are high and river flows are already low. This has resulted in abstraction restrictions in several areas of the UK during the main irrigation demand period for potatoes. To overcome this problem winter abstraction from rivers and on-farm water storage has increased rapidly over the last 5 years. This approach secures the grower's supply for the following season and avoids the need for abstraction during the summer when river flows are normally low. There is however an increased cost to the grower associated with the construction of a reservoir, but increasingly this is regarded as worthwhile, as water supplies can then be guaranteed. Low flows in rivers in summer will necessitate the construction of reservoirs in Finland as well.

Although there are a number of irrigation scheduling techniques and services available (Chapter 5.3) the number of growers using them remains small. Estimates suggest that as few as 15-20% of growers use a recognised service. In Portugal where it is estimated that around half the water used in agriculture is used inefficiently, representing a 50% loss (EEA/UNEP, 1997),

Table 1   Irrigation practice in Italy.

Region	Irrigation method		Scheduling services
	Past and 80's	Current	
Piemonte Lombardia Veneto	Furrow Irrigation Rain gun, high press.	Rain gun, medium pressure (30%) Rain gun, high pressure (65%) Irrig. booms (5%)	Available but scarcely utilised
Emilia Romagna (Bologna)	Rain gun (medium/low press.)	Irrig. booms (50%) rain gun (50%)	Available, 40% of the growers access it
Campania	Furrow irrigation	Furrow irrigation (80%) Rain gun, medium/low press. (20%)	Not available
Puglia	Furrow irrigation Rain gun	Drip / trickle irrigation (70%)+ plast.mulches (several) Irrig. booms (10%) Rain gun medium/low pressure (10%) Furrow irrigation (10%)	Not available
Sicilia Sardegna	Furrow irrigation Rain gun	Rain gun medium/low pressure 65% Drip / trickle irrigation (25%)+ plast.mulches (several) Furrow irrigation (10%)	Not available

apparently no scheduling services are available. Table 1 illustrates the use of scheduling services in Italy as well as the development of irrigation techniques. There may be a number of reasons for this low usage. Firstly, although the cost of many scheduling systems are small compared with their potential benefits, some growers may be quickly discouraged if they pay for scheduling only to find that rainfall minimises the need for irrigation. Secondly, most growers will have insufficient equipment to follow all the advice given by the scheduling service during particularly dry weather. Thirdly, some growers still believe that local experience over a number of years is all that is required. In some areas there are no extension services promoting the use of irrigation scheduling. No doubt if the cost of water increased appreciably the need to use water more efficiently would encourage growers to use a scheduling service.

If water supplies for irrigation are restricted growers need to decide whether the remaining supplies should be used to achieve the desired quality of produce or whether increased yield will give the best return. The circumstances of the individual farmer and the markets they are growing for will determine which option is taken. A recent EU (1998) report stated *'The scope for research and technological development is extremely broad. A number of technologies and management support tools have been developed but these are not yet widely applied either because they are not easily implemented in the context of the existing agricultural settings (and would therefore require further research for adaptation) or because their applicability has not been demonstrated under the variety of agro-climatic and socio-technical conditions prevailing in the EU'.* There remains much to be done in the promotion of good irrigation practice.

The most common method of irrigation in Europe is the rain gun. This technique is relatively inefficient and inaccurate but it is cheap and mobile and offers considerable flexibility. However, in recent years the demand for consistently high quality produce has led many growers to change to irrigation booms which are capable of uniform applications of water even in relatively windy conditions. Trickle or drip irrigation is known to be very accurate and may reduce both aerial and soil surface evaporation losses thus increasing efficiency. It is used in many parts of Europe but cost and problems retrieving the tape or pipes from the field at the end of the season has prevented its more widespread adoption. If water resource allocation for agriculture is reduced the use of this technique will increase, possibly in conjunction with fertigation which will possibly allow better use of nitrogen as applications can be better linked to crop need.

Nitrogen

In Europe many countries are experiencing increasing concentrations of nitrate in public drinking water supplies (EEC, 1991). Levels commonly exceed EC guidelines. Eutrophication of natural waters is linked to increasing concentrations of nitrate (in some areas phosphate as well).

The potato crop is grown extensively in soils overlying aquifers supplying water for domestic consumption. Where the potato crop receives superoptimal rates of nitrogen the levels of potentially leachable nitrate post harvest of the crop can increase dramatically (Fig. 2). However what might be regarded as the optimum economic N rate for yield of standard ware sized potatoes might not be sufficient if large tubers are required. It has been shown (Goodlass & Johnson, 1997) that for maximum yield of tubers in the 65-85 mm range of tubers N rates well above those given in standard recommendations are required. The calculated optima reported were commonly above the highest rate of N used in the experiments (350 kg/ha). These levels will be above the biological optima which appears to be the breakpoint in the response of residual soil mineral nitrogen levels to increasing levels of nitrogen application. Where small potatoes are required the economic optima will be lower than that generally recommended.

Many potato crops receive super-optimal N fertiliser rates particularly where organic manures are used. Growers have little confidence in the suggested supplies of N from organic manures given in standard recommendations. They

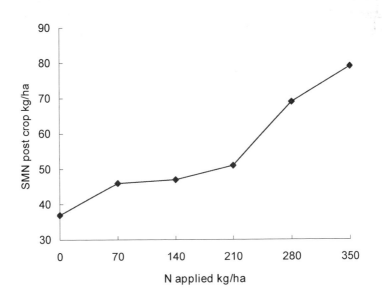

Figure 2   Residual soil mineral nitrogen after potatoes (Biological Nitrogen optimum was 256 kg/ha) SMN = Soil Mineral Nitrogen.

are express concern about the variability in analysis of manures (particularly slurries) and about their own ability to spread them evenly at a known rate. Growers have very little appreciation of losses of ammonia from applied organic manures and the need for rapid incorporation if these are to be reduced. It is ironic that applications of dilute slurries (Fig. 3) result in less ammonia loss but require a greater energy input as more water has to be moved around the farm. Research in Italy in 1997 indicated that growers who used FYM made some adjustment to their applications of N but 100% of those applying slurries did not. Many growers use more N than is suggested in standard recommendation systems, applications in some parts of Italy (Table 2) seem very high, especially if they have also used an organic manure. Table 3 indicates that reductions in levels of N fertiliser applied to potato crops in the UK where organic manures are applied is less than where cereals or grass are grown reflecting the 'insurance' approach to N fertilisation of potatoes by UK growers.

The decision support system MANNER (MANure and Nitrogen Evaluation Routine) (Chambers et al., 1999) available to growers in the UK to assist with decisions on residual nitrogen levels following the application of organic manures needs demonstrating in local situations before the majority of farmers will have faith in it. Even when systems such as these are used as an aid to N

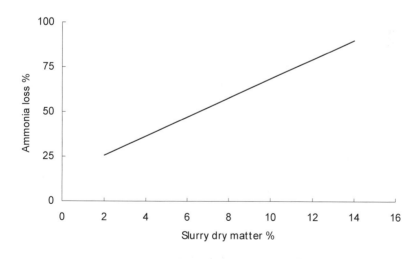

Figure 3   Ammonia loss from surface applied cow slurry.

Table 2   *Nitrogen use on potatoes in Italy.*

Region	Applied mineral N kg ha^{-1}		
	Past and 80's	Current	Split applications (current)
Piemonte Lombardia Veneto	150	200	Yes (at planting + side-dress applications)
Emilia Romagna (Bologna)	250	200	Yes
Lazio	150	250	Yes
Campania		350	Seldom
Puglia		300	Yes fertigation (several farms)
Sicilia Sardegna		350	Seldom

prediction there are weaknesses. If a grower uses MANNER for example in January before ordering his fertiliser he takes the risk that rainfall will be higher or lower than average and thus leach more or less N out of the soil, unless he splits his nitrogen applications he will have under or over ordered N fertiliser. Where crops are irrigated split applications of N are tool for the grower using

Table 3   *Reduction in N fertiliser rates on crops receiving organic manures compared with those which do not.*

Crop	Area spread with organic manure (%)	Reduction in Nitrogen applied (kg/ha)
Winter wheat	11	22
Maincrop potatoes	35	4
Grass for silage	66	11

Weighted mean data from British Survey of Fertiliser Practice 1988 - 1993 organic manures. Unirrigated potatoes rarely receive split N applications and all decisions on N rate are taken pre-planting, as potatoes are a high financial return crop and the grower knows that the financial penalty for over application of N is small he will err on the side of caution no matter how many times the scientists tell him that applying 50 kg/ha N less that optimum will have only a marginal effect on yield.

Unless forced by agreement or legislation farmers are reluctant to make allowances for the N contents of organic manures which are often applied at excessive rates prior to the potato crop. The implementation of the EC Nitrate Directive will reinforce local rules on manure applications (as in the Netherlands and Flanders) and in others introduce rules which demand that the contribution of N from organic manures is taken into account when growers decide on their rates of artificial manures. At some stage a decision will need to be made as to which causes the most environmental harm, nitrate loss following the use of manures which have been incorporated or injected into the soil or ammonia loss following the surface application of manures.

Within the Nitrate Sensitive Areas scheme in England the largest concentrations of nitrate in water draining from land following potato crops (Fig. 4) were where organic manures had been applied (sites B,L,M,N) or where drought restricted

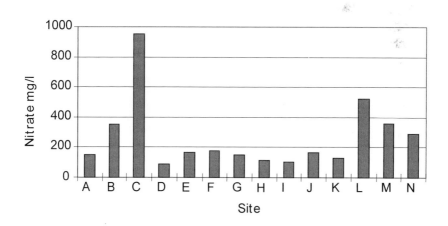

Figure 4   Nitrate concentrations in drainage water following potato crops grown in Nitrate Sensitive Areas in eastern England.

the growth of the crop (site C). Losses of nitrate before the potato crop can be reduced by the use of cover crops but the speed of mineralisation of the residues from the ploughed in cover crop is then an uncertainty in predicting the N fertiliser requirement of the potatoes. Post crop losses can be reduced in some years by growing a winter cereal but in others delaying cultivation until the January following harvest of the potato crop will reduce nitrate concentrations in drainage water (Shepherd & Lord, 1996).

Potatoes are only one of many crops grown in rotations and if nitrate losses to water are to be reduced there is a need for action with the whole rotation. Results from the Nitrate Sensitive Areas scheme in England (Fig. 5, Lord et al., 1999) show that reductions are possible with many crops. The dramatic change in concentrations following potato crops was closely linked to improvements in the use of nitrogen from organic manures.

Nitrate loss can also occur during the early part of the growing season, particularly on sandy or shallow soils, if rainfall is high and if mineralisation rates are high (Neeteson et al., 1989). Most growers have discovered that this results in a loss of yield and therefore they split their nitrogen applications, commonly supplying half their nitrogen at planting and half at tuber initiation.

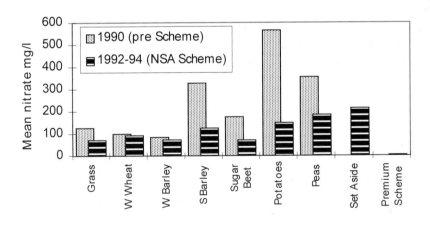

Figure 5   *Mean drainage water nitrate concentrations pre and post Nitrate Sensitive Area scheme introduction in England. ('Premium Scheme' is unfertilised ungrazed grassland.).*

The use of fertilisers containing nitrification inhibitors, such as dicyandiamide (DCD), has been suggested as an alternative to this approach. There is some evidence (Bailey et al., 1992) that this approach can work on a sandy soil in some years but there is a need for more experience with this material on other soils as a simple two way split has been shown to give the best results (in yield) on calcareous shallow soils.

Effect of the EC Nitrate Directive

The implementation of the Nitrate Directive will mean that growers in large areas of Europe may need to justify the amounts of nitrogen they apply to their potato crops. The attitudes of the regulatory authorities could have a significant effect, will they accept standard recommendations or will they require a more detailed examination of supplies of N from the soil and particularly organic manures? If they take the later option there should be an increase in the use of soil mineral nitrogen sampling and also, because of uncertainties over the rates of mineralisation of N from residues and manures, an increased need for plant sampling during the growing season combined with split applications of N. There is therefore a particular need for improved models to assess N need from plant analysis based on chemical tests or non-invasive techniques as described earlier in this handbook. These tests will need to be cheap and possibly tractor mounted so that variable rates of N can be applied across the field.

Conclusions

Water is a precious commodity even in the maritime areas of Europe and must therefore be used efficiently. This means that more use should be made of the scheduling systems available to potato growers, otherwise they will find that as water resources come under more pressure from industrial, domestic and environmental users they will be restricted in the amounts they can use. This is already happening with nitrogen fertiliser. The efficient use of nitrogen from organic manures is already demanded by the EEC Nitrate Directive, but farmers in general have little faith in the standard approaches to the estimation of nitrogen supplies from manures - or is it that the potato crop is so vital to their economic well being that they use an insurance approach? Unless scientists and extension workers can produce 'user friendly' and efficient decision support systems for both water and nitrogen a continued inefficient use of these resources could lead to more legislation. Growers already have to cope with legislation on many of their activities and to retain their faith in science we must produce predictive systems which they find accurate.

Recommendations

1. Irrigation scheduling services should be available to all farmers at a reasonable cost. Their benefits must be promoted.
2. Extension workers should be trained in the use of nitrogen prediction techniques (possibly all extension workers dealing with fertiliser should be able to show that they are adequately trained). A qualification such as that of FACTS (Fertiliser Advisors Certification and Training Scheme) already in operation in the UK combined with a necessity to show ongoing training is one possibility.
3. There is a continuing need to produce 'user friendly' nitrogen prediction systems. Specifically those which can be carried out during the growth of the crop.
4. Potato growers must be encouraged to make full use of the nutrient contents of organic manures (not just the nitrogen) through expert systems such as MANNER and by the use of demonstration farms.

6.2   Precision farming for the management of variability

B. Marshall

What is precision farming - a buzz phrase of the 1990's or something central to European farming of the future? What are the potential economic and environmental benefits on offer? Why now? What are the tools that are needed? What is GPS? Can sources of variability in yield and quality be better quantified and managed? What is the role of decision support systems? What are the challenges in implementing precision farming for the better utilisation and conservation of water and nitrogen resources in potato production? What is still to be done?

Introduction

Precision farming is the active management of variability rather than simply 'living with variability'. Its purpose to improve efficiency by reducing wastage, reducing environmental risk and improving gross margin. This is achieved by precise monitoring, by improved understanding of the interactions between environment (aerial and soil), agronomic inputs and yield, and by varying treatments both spatially and temporally both between and within fields to match the local conditions. It is enabled by modifying current techniques, incorporating new techniques and the use of computer based decision support systems. Precision farming also implies a sustainable agricultural industry at several scales both spatial and temporal - e.g. there should not be a net-loss of resource from the farm enterprise nor a build up of pollutants and waste. On a wider scale the resources imported to the farm should be renewable and the waste should ultimately be recyclable. At the same time the farm also has to be economically viable from year to year. This involves risk management.

The word 'precision' refers strictly to variability. However, in the context of precision farming it also incorporates the important concept of improved accuracy, more accurate targeting of inputs to match expected output. There is no point in doing the wrong thing precisely!

Knowledge of the variability of yields over seasons (temporal variability) was and is important in the farmer's strategic planning of economic risk. The farmer will also have knowledge of the variability of yield levels between fields on the farm. However, until very recently, detailed recording of the spatial variability of yield within a field was not a practical proposition. Therefore, decision making has been, and continues to be, based on field averages of yield. Inevitably this means that some areas receive supra-optimal and others sub-optimal applications of inputs. Harvesting machinery fitted with GPS can provide maps of the yield variation within fields. Farmers may also have additional spatial information, albeit at a cruder resolution, on aspects such as soil fertility, soil moisture properties etc. which influence yield. The potential benefits of having such information are poorly quantified.

Causes of variability

Variability occurs in two forms: spatial and temporal. Table 1 lists the majority of causes of spatial variation in yield, mostly associated with below ground characters. In addition, the table shows which characters also exhibit significant temporal variation and includes above ground characters that show only temporal variation. Some characters express their effects through others e.g. clay content through its effects on soil moisture and chemistry. Characters showing spatial but not temporal variation should produce patterns which are fixed throughout the season and consistent from one season to the next. These characters should be easier to diagnose as causes of variation than those subject to temporal variation that may exhibit only transient patterns. Unfortunately many characters associated with water and nitrogen vary both spatially and temporally.

Table 1   Possible causes of yield variation within a field. A star indicates that the character
is likely to produce significant variation within a field (spatial) or significant temporal
variation, from season to season or more rapidly.

Cause	Spatial	Temporal
Below ground		
moisture content	*	*
ground water level	*	*
temperature	*	*
fertility (N)	*	*
fertility (K, P, Mg)	*	
organic matter content	*	
texture, type, depth	*	
clay content	*	
pH	*	
pests & diseases	*	*
drains	*	
compacted layers	*	
runoff	*	*
Above ground		
pests & diseases	*	*
weeds	*	*
shading at field margins	*	*
solar radiation		*
temperature		*
humidity		*
wind speed & turbulence	*	*

Potential benefits

There is limited information on the magnitudes of the spatial variability of yield
occurring within fields in any crops. Van Kraalingen (1997) has attempted to
estimate the scale of variability within the field and the potential benefits of
managing this variability by spatially varying the rates at which inputs are
applied to the crop. Observations on winter wheat crops (typical spatial

resolution of samples being 10 m^2) suggest that a coefficient of variation of 15 per cent would be a conservative estimate of the variability i.e. the standard deviation of local yields being 1.5 t/ha in a crop yielding 10 t/ha averaged over the entire field. In potato crops, this variability is likely to be greater. For example, only specific ranges of tuber size are acceptable for particular markets. Tuber size is inherently variable within the potato plant and any slight changes in local yield could have large effects on the proportions of saleable product. In contrast, cereals have a much larger range of acceptable grain sizes relative to their intrinsic variability. The potato crop has a shallower rooting depth than winter cereals especially and is therefore more susceptible to shortages of water supply in the upper soil layer. Even taking the conservative value of 15 per cent and assuming that with the implementation of precision farming all yields below the mean are brought up to the average yield, not the maximum observed in the field, van Kraalingen estimated that net profits would be increased by more than 10 per cent. While herbicide input was assumed to be reduced by one half, it was assumed that quantities of all other inputs remained the same but were more timely in their application and or more accurately targeted to local variations within the field. The higher degree of variability likely in potatoes and the opportunity for reducing inputs suggest the potential benefits should be even greater in both economic and environmental terms.

Measurement of spatial variability

Yield

The first, essential step in implementing the spatial aspects of 'precision farming' is to quantify the variability. This requires an inexpensive and convenient system for mapping yields. Only then can the economic and environmental benefits of spatially variable treatments be assessed. Various techniques for positioning have been tried but the recent incorporation of global positioning systems (GPS) and real-time weighing into harvesting machinery offers the most practical method of recording the spatial variations of yield within fields (Stafford, 1998). GPS uses 24 satellites owned and operated by the US Department of Defence, which permits civilian use but at a much lower resolution (typically 100 m) than that for military purposes. This resolution is too low for agricultural use. A higher resolution is achieved by using two separate receivers. One receiver, sometimes referred to as the base station, is placed at a fixed, known location on the farm. It estimates its position from the satellites (a minimum of four must be in view of the receiver at any one time), compares the estimated position with its true position and transmits the positioning error to the second receiver. This second, roving receiver then applies this error correction to its own estimated position. This dual receiver system is known as differential GPS (DGPS) and achieves a much better resolution of 2-5 m. Figures 1 and 2 give examples of the precision and accuracy to which a vehicle's (e.g. tractor, harvester) location can be estimated by DGPS. Figure 2 also illustrates a potential difficulty due to loss of signal. While the resolution is adequate for yield mapping, fertiliser application and irrigation (typically requiring 10 – 30 m

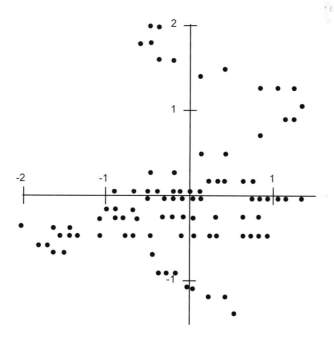

Figure 1   Static precision of a differential GPS receiver where both the receivers, one mounted on a tractor and the other at the local base station, were kept in fixed positions. Individual point locations are estimated at 2-second intervals. Axes scales are deviations in m from the mean location averaged over the two-minute interval. The figure is redrawn from Stafford (1998).

resolution) it may be too coarse for avoiding spray overlap, row crop planting etc. (typically 5 - 100 cm resolution). New systems are under development, based on kinematic GPS, which can achieve this finer resolution (Stafford, 1998).

Harvesters fitted with DGPS devices for combinable crops are commercially available. Such harvesters for root crops (potato, sugar beet etc.) are only in the early development phase (Campbell, 1998). There are substantial technical difficulties in estimating fresh weight yields in real time due to the variable amounts of soil which adhere to the tubers. There are also difficulties introduced by the physical displacement and mixing of tubers during the harvesting. As the tubers are lifted they are passed backwards along the web at a rate which is in generally faster than the forward speed of the tractor. The net effect is that the tubers are deposited into the collecting trailer behind the point that they were lifted from. The extent of this backward displacement depends on the speed of the web which is adjusted according to soil conditions at the time. More difficult to resolve, is the problem of roll-back. While passing up the web some tubers roll back and mix with more recently lifted tubers. This produces a form of spatial averaging which will limit the spatial resolution attainable. The effect will vary with soil conditions, slope within the field and may be preferentially greater with larger than smaller tubers. Clearly these effects need to be quantified and taken into account.

Soil characters

In principle any character can be spatially sampled and mapped. The main constraints on such sampling are the cost and time required for collection and analysis. Currently pH, P, K and Mg are mapped by some growers. These are characters that do not change appreciably in time (Table 1). Soil type, depths of soil horizons of moisture holding capacity and organic matter content would also appear to be useful characters to be mapped. All these characters have a bearing on the irrigation and nitrogen fertiliser requirements of the crop either directly through their contribution to water and nitrogen supplies or indirectly by determining the local maximum yield level and hence local water and nitrogen demand. Although the locations of the samples are known accurately their resolution (typically 100 m grid) is considerably less than that of either the

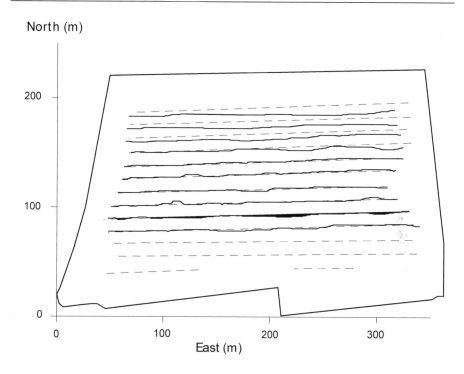

Figure 2   *Dynamic accuracy of a differential GPS receiver. The system was the same as in Figure 1 except that this time the tractor with receiver was tracking (solid lines) tramlines (dashed lines). The tramlines had been accurately surveyed previously by conventional means. The poorer resolution to the north of the field was possibly caused by an avenue of trees obscuring the signal. The figure is redrawn from Stafford (1998).*

yield maps or machinery which can apply inputs at variable rates. Access for sampling also presents difficulties during the growing season. Thus most field sampling for mapping purposes is best done before planting and at or after lifting. Unfortunately, both mineral nitrogen content (see section 3.4) and soil moisture (see sections 4.3) are highly variable in time and hence do not lend themselves to direct measurement and mapping, especially during the season.

Remote sensing

Remote sensing offers a faster method of mapping some characters, or surrogates for them, at spatial resolutions similar to or finer than that of the farm machinery. Such characters include Leaf Area Index (LAI) or ground cover, stress indicators (canopy temperature, moisture status of the crop (Cerovic *et al.*, 1996), N status of leaves - see also section 2.3 for individual leaf assessment), soil surface temperature, soil reflectance, weed distribution and intensity. The use of these measures are still being researched or developed to assess their robustness in practice. However, sequential measures of ground cover by remote sensing have already proved reliable in estimating sugar beet yields (Clevers & Van Leeuwen, 1996).

Management of variability

An explosion of interest in precision farming led to conferences and research papers. One conference in the USA alone ran to 60 papers and 100 abstracts. In the proceedings of that conference there was a forward looking paper by Rawlins (1996) entitled '*Moving from precision to prescription farming: The next plateau*' in which he observed,

> 'The emergence of a number of technologies at affordable prices, has made it possible to precisely apply spatially-variable inputs to farmers' fields. At present the prescriptions for these inputs are typically empirical, based primarily on grid sampling and soil tests. Such prescriptions draw only on a small fraction of existing scientific knowledge of processes controlling crop growth. These empirical prescriptions work well for P, K, Lime and other inputs that don't leach or volatilize, but the primary variables controlling crop yield are more often water, nitrogen, pest or diseases or other factors that require within season management. Developing prescriptions for real-time management of these inputs will require the use of far more knowledge about the processes limiting crop yield at any time at any area of the field than we're now using. Crop simulation models provide a means to package a huge reservoir of scientific knowledge about these processes in a form that can used for prescribing inputs. But to be used, these simulators will need to be validated in the field, and will require much denser site-specific

environmental data sets to feed them than we now have. Fortunately, the technologies to bridge these gaps appear to be here or on the near horizon. But a coordinated effort of a team composed of government, academia and industry members will be required to move to this next plateau'

Pest and diseases in the potato crop should be under adequate control, leaving water and nitrogen, the focus of this concerted action, as the two key primary inputs determining variability in crop yield. Both nitrogen and water can be applied at spatially variable rates. Of the two, technology for applying nitrogen at variable rates is cheaper and is easily moved (tractor mounted) from one crop to the next. The technology is applicable to the whole of European agriculture. In contrast, the application of site-specific irrigation requires equipment dedicated to specific fields (linear booms or centre pivot irrigation under computer control). Fertiliser can be applied through the same equipment. The use of this relatively expensive equipment is most likely in the drier regions of Europe where returns on investment are greatest. Even without this equipment, the question still remains as to whether to irrigate or fertilise according to the lowest, the average or some other yield level, bearing in mind the non-linear response of yield to the amounts of irrigation and nitrogen applied and the interaction between them.

The first stage in managing spatial variability is to diagnose the causes of yield variation correctly. This exercise is far from trivial, even in cereal crops grown continuously in the same field. Since potato is grown in a rotation, there is no corresponding yield map from the previous year and a map of potato yield could be four or more years old. It might be possible to use a yield map from the crop immediately preceding potato in the rotation. As yet, systematic studies relating yield variation in one year to those in next, even in the same crop, are limited. Assuming the causes of variation can be diagnosed correctly, the economic and environmental benefits of alternative management strategies can be evaluated.

A sophisticated Decision Support System will be needed for the diagnosis of cause and in the selection and evaluation of appropriate remedial action. The cause of yield variation may change both in space and time. Variations in available soil moisture due to changing soil depth may produce spatially variable symptoms of stress that are manifest strongly in dry years but not in wet ones. That is, the pattern of symptoms would appear to change from one year to the next even though the underlying pattern of cause (soil depth) remained constant. A model of water limited yield would be essential in testing whether such changing patterns of symptoms were consistent with the expected changes in water supply. Crop models that take account of the physical and chemical environment will be important in both diagnosis and evaluation (Chapter 5). Rule-based techniques from the Artificial Intelligence discipline provide a useful framework for developing diagnostic aids. Uncertainties in several aspects (e.g. sampling error, diagnosis, future weather and crop response) require the introduction of risk assessment in the selection of appropriate remedial action. Bayesian belief networks provide a tool to handle and propagate such uncertainties.

Figure 3 shows one schematic of how a decision on the optimum strategy for application of water and nitrogen fertiliser could be reached. Starting with no applied water (neither irrigation nor rainfall) and no applied fertiliser N, water and nitrogen are 'applied' at each location in the field in just sufficient quantities to remove any local limitation on yield. These requirements are calculated using appropriate crop models (Chapter 5) and produce two spatial maps of the requirements for total amounts of water and nitrogen to be applied during the season. In practice, water is applied uniformly over the entire field, whether as rainfall or irrigation. So, a uniform rate of irrigation has to be selected which may be sub-optimal in some areas and supra-optimal in others. The sub-optimal areas lead to reduced yield and thus reduced nitrogen demand for growth locally. The spatial map of nitrogen requirement is adjusted accordingly. Supra-optimal areas have increased risks of nitrogen leaching. Thus an optimum compromise between the risks of economic loss (yield) and environmental damage (leaching of nitrogen) can be determined. The result is a water requirement (rainfall plus irrigation) to be applied uniformly and a nitrogen requirement which is spatially variable. The purpose of this exercise was to illustrate the complexity of such a decision making process and the need for robust models of water supply, soil nitrogen and crop growth for the benefits of precision farming to be realised.

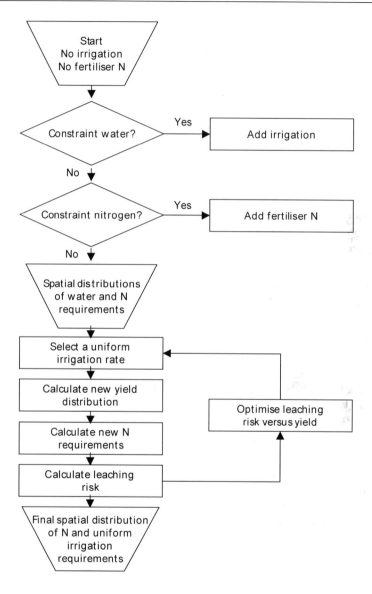

Figure 3 Schema for decision making on irrigation requirement (uniform) and spatially variable application of fertiliser N in relation to precision farming. Irrigation requirement includes both applied water and rainfall.

Fertiliser application

The technology to apply variable rates of inorganic fertilisers and urea is available and in use on some farms. The technology to measure nitrogen content of slurry as it is being applied is being tested now. Thus slurries could be applied at varying rates or the data on nitrogen applied could be stored and a supplementary dressing of inorganic fertiliser applied at a spatially variable rate afterwards. The development of real-time sensors for solid manures presents considerable and possibly insurmountable technical problems. Alternative approaches, when using solid manures, would be intensive sampling of the soil prior to planting or using the crop as an indicator of nitrogen requirement. Intensive soil sampling to produce detailed maps of soil mineral nitrogen and organic matter content is unlikely on economic and feasibility grounds. Using the crop as an indicator from ground based samples is also likely to be too time consuming and expensive, sufficient only for treating the crop as a uniform average, not as spatially variable. Remote sensing would appear to be the only realistic means of spatially sampling the crop status quickly enough, with the opportunity of recording several such maps at key stages of crop development.

Perhaps the most immediate opportunity to exploit spatial mapping is when the potatoes are harvested. The resultant yield map indicates areas of higher yield and potentially larger amounts of crop residue . This information could be used to reduce nitrogen inputs in the following crops in those areas of high residue. Correct interpretation of symptoms and their causes will be crucial to the successful use of such data.

Conclusions

- Precision farming is the active management of variability; the accurate and precise targeting of inputs to match outputs both in space and time.
- A conservative estimate is that net profits could be increased by more than 10 % by bringing the lowest yielding areas of a field up to the average for the field.
- Precision farming is a new name but not a new concept. It has never been in the growers interest to waste resources.
- The degree of precision is rapidly improving for several reasons:
  Technology

- Experiment and crop modelling has led to a better, quantitative understanding of how fluctuations in nitrogen and water supply cause variations in crop yields.
- The increasing accuracy and precision with which inputs, particularly fertiliser, can be applied.
- The advent of GPS, allowing spatially referenced data on soil characters, topology and crop performance to be collated rapidly.

Incentives
- The increased awareness of risk to the environment is also demanding greater accuracy and precision in the application of fertiliser.
- Limited water resources demand greater precision.

- Major challenges for precision agriculture
  - Continued improvement of real-time monitoring of inputs
  - More efficient means of remotely sensing crop status with adequate spatial resolution.
  - Development of robust decision support systems that (i) can combine knowledge from a range of sources which vary in their spatial and temporal resolution, (ii) aid diagnosis of yield variation (iii) handle uncertainties in the economic and environmental evaluation of alternative management strategies.

Acknowledgements

I would like to thank Paddy Johnson (ADAS, UK), and Remmie Booij (AB-DLO, NL) for their constructive comments.

6.3   Organic potato farming and nitrogen fertilisation, some food for thought

R.E. Wheatley, N.U. Haase, W. Cormack, M. Colauzzi & G. Guarda

Acceptance of organic farming has increased during the last few years, but several questions still need answering. For example is organic cultivation an environmentally friendly and effective agronomic system in which all resources are used effectively or are some wasted? To this end how do we ensure that the nutrients are available to the crop at the correct time, and that mistiming does not cause either environmental damage or reduction in tuber yield and quality? We also need to know what changes in management practices are required to make organic production economically sustainable.
Growers need to be 'biologists' and empathise with the whole system in an 'holistic' approach.

Introduction

The philosophy of organic farming requires closed systems with the minimal import of nutrients. These systems should be sustainable and give a high quality product without the use of manufactured chemical fertilisers or biocides (Van Delden, 1988). Manure is used to 'feed the soil' and so indirectly the plants. The sources of these manures are restricted. The biggest single limitation to potato yield in an organic rotation is likely to be the supply of nitrogen (Cormack, 1997). The aim of this chapter is to review the whole question of organic cultivation of potatoes, particularly with reference to the supply of nutrients such as nitrogen.

Crop Nutrition

Manures are used in many organic farming systems as a source of nitrogen, phosphorus, potassium and micronutrients such as trace metals. However nitrogen can leak from the system but phosphorus is not so mobile and so may accumulate over a number of crop cycles. Saturation of the soil with phosphorus may result in leakage to the ground water causing eutrophication. In such cases it will be necessary to add nitrogen to the soil without any accompanying phosphorus. This can be achieved by using leguminous crops in rotations. The nitrogen-rich plant residues are incorporated into the soil, and subsequently mineralised to provide soil nitrogen. Only a small amount of mineral-nitrogen leaks from the nodules during growth.

Table 1   Examples of crop rotations that include potatoes (after Kölsch and Stöppler, 1990).

Year	a	b	c	d	e	f
1	clover grass	red clover	red clover	red clover	clover grass	clover-lucerne
	self destroying	-	-	-	-	-
2	**POTATOES**	oat/winter rye	winter wheat	oat	**POTATOES**	clover
followed by	-	peas/lupine	peas	-	-	-
3	winter wheat	**POTATOES**	**POTATOES**	winter rye	winter-rye	oat
followed by		-	-	lupine	rape seed	-
4	winter- rye	winter rye	carrots	**POTATOES**	faba beans	winter wheat
followed by	clover-grass	-	-	-	-	legumes
5		winter barley	winter wheat	winter rye	winter wheat	**POTATOES**
followed by		red clover	peas	serradella	-	-
6			summer barley	summer barley	winter barley	winter rye
followed by			red clover		clover grass	legumes
7						summer barley
followed by						clover-lucerne

Such a rotation could start with red clover, which is then cut and left as mulch before ploughing-in in April of the following year, prior to the planting of potatoes. Wheat is sown in the third year, beans in the fourth, and wheat again in the fifth, before returning to red-clover in year six, Table 1. There are also many other possibilities. The right choice of crop rotation is important for effective production.

Estimating how much nitrogen will be introduced into the soil from the breakdown of these residues and at what time presents major problems. Both of these are difficult to predict as soil mineral nitrogen concentrations can vary greatly throughout the season, Figure 1.

Timing is critical. Cultivation must be timed so that the subsequent mineralisation (Smith & Chalmers, 1993) will provide nitrogen when the potatoes require it. Potato plants take up between 50 % and 70 % of their required nutrients within 45 days of emergence. Failure in the nutrient supply at this critical time will result in a lower yield. Late mineralisation to available nitrogen forms during the growing season can be a problem. This may affect tuber maturity and quality and also have an adverse affect on storage potential (Müller, 1997). These hazards are common to both organic and conventional farming.

The difficult logistics involved in the control and estimation of the time and amount of release of ammonium-nitrogen from soil organic matter, and its subsequent conversion to mobile nitrate-nitrogen, that potentially can be leached into the groundwater or reduced to greenhouse gases, are common to all agrosystems. The mineralisation efficiency of soils depends on several factors, such as climate, soil type, management history and crop activity. Cultivation in autumn, leaving the soil bare over winter will result in very significant losses and significant environmental damage, from both leaching and denitrification.

Problems of losses, particularly via leaching, will vary according to the soil type. Losses from 'retentive' soils that contain significant amounts of clay will be significantly less than those from free-draining soils such as sands, which also

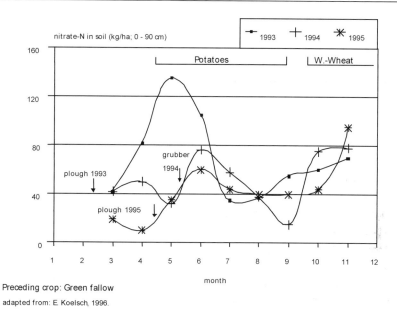

Figure 1   *Nitrate dynamics in soil (0 – 90 cm depth) under potatoes, following winter wheat preceded by a green fallow crop.*

have a low exchange capacity. The incorporation of organic matter into such free-draining soils may assist the situation by increasing water-holding capacity. Under present EU regulations almost any organic matter can be used. However this may conflict with the 'organic farmers' philosophy that all systems should be sustainable.

Present and future controls on the disposal of municipal sewage will result in the production of large amounts of material that have to be disposed off and incorporation into soil is an obvious, if not entirely desirable, answer

How effective is 'organic cultivation' in providing nutrition for potatoes

Potato yields in a five-year rotation sequence that includes two nitrogen-fixing crops can approach those from 'conventionally' inorganically fertilised crops.

Soil type is again very important and the highest yields will be achieved on soils such as silty-clay loams that have a high exchange capacity. Nutrients are held in such systems and leaching losses are low. A major reservation to sustainability in such systems is the ability to maintain sufficiently high potassium and phosphorous levels in the soil. Other sources of potassium and phosphorous include seaweed, which has been used traditionally for many years by coastal communities and other materials such as composted bark, wood-ash, basic slag, and rock-phosphate. Basic slag is a waste product of the iron and steel industry, and so is in plentiful supply. Problems associated with slow release from these sources can be overcome by adding them to the system when legumes are sown.

So good farming practice, including the right choice of crop rotation is required. Organic fertilisers such as farmyard manure can be added to rapidly increase nutrient levels when necessary (Table 2).

Table 2   *Effect of manure on potato yield (t/ha, averages of 9 varieties) (Kölsch and Stöppler, 1990)*

Application	Neu-Eichenberg		Ellershausen	
	1984	1985	1984	1985
No manure	25.7	18.5	21.6	15.9
20 t/ha (~80 kg N/ha)	26.7	20.3	22.6	15.5
40 t/ha (~160 Kg N/ha)	28.4	20.2	23.6	16.8
LSD 5%	1.2	0.9	0.9	0.9

Economic and Environmental Implications

Production levels are more frequently about 50 - 60% of the yield from conventional farming (Table 3), of which a large proportion is sold directly to the consumer at the farm. However more supermarkets and manufacturers of potato products are buying these potatoes and distribution systems are evolving. Potatoes produced by organic farming methods attract a premium price. Consumers are willing to pay the added costs associated with lower yields in what is perceived to be an environmentally friendly growing system!

Table 3   *Potato yields from conventional and organic farming in Germany) (adapted from Neubauer, 1997).*

Year	conventional (t/ha)	organic (t/ha)	organic to conv. (%)
1990	28.9	16.1	56
1991	27.4	17.3	63
1992	30.9	14.5	47
1993	32.4	17.1	53
1994	28.0	19.2	69
mean	29.5	16.8	57

Today the high price levels for the lower yields result in an acceptable income for farmers.

Optimisation of productivity in organic farming

Evaluation of organic manures

Several billion ECU's worth of nutrients are added to crops via organic manures annually in the EU. The efficiency of use is dependent on the specific nutrient. In most situations both phosphorous and potassium are used reasonably efficiently but nitrogen use efficiency can be reduced by poor management practice.

Measurement of total nitrogen

It is possible to measure or calculate the nitrogen that is in manure in several ways, e.g.; Dumas combustion for total N, Kjeldahl for organic N and spectroscopy for inorganic N (Chapter 2.2). Total nitrogen in manures is generally between 0.5 and 0.75 % of the fresh weight. About one third of this nitrogen is present as ammonium-nitrogen. Between 7 and 80% of this ammonium-nitrogen can be lost by volatilisation if incorrectly applied to the field. Using the correct management practices can reduce such losses. For example, application when the soil is cool and moist, followed by rapid incorporation.

Measurement of mineral-nitrogen

Ammonium-nitrogen levels in slurry can be measured 'on-line' during application to the field. However it is better practice, both economically and environmentally, to measure the total nitrogen before application and use a computer programme to estimate total nitrogen application rates. Also although the total nitrogen content of the manure can be readily estimated, timing of availability to the crop is critical.

Measurement of the carbon:nitrogen ratio

Carbon to nitrogen ratios have to be measured because a high ratio will inhibit mineralisation in the manures. Long term organic farming will increase the carbon content and reduce the carbon to nitrogen ratio in the soil.

Other major nutrients in the manures

As animal manures provide the major source of phosphorous and potassium in many organic systems, accurate analyses for these elements is essential. A major decision lies in the choice of extractants, which can range from water to mineral acids. The extraction methods used will affect the analytical results as these elements can be in different forms in the soil and soil organic matter (Chapter 3.2).

Time of application of manures and cultivation

To avoid structural damage to clay soils manures should be applied at the end of the growing season. Lighter, more-freely draining soils may be more assessable over a longer period, and so application throughout the winter and at the start of the season may be possible. However great care must be taken to ensure that the nitrogen released from such applications is taken up by plants and not leached away in the ground water.

So timing of application together with the use of either green manures on autumn cultivated soil or non-disturbance until spring followed by minimal cultivation to incorporate the manure and plant the potatoes, is extremely critical.

Spring applications of pig and cattle slurries and poultry manures to crops are generally more efficient than autumn applications, particularly on freely draining soils. And spring cultivation to incorporate leguminous residues will also be preferable to autumn cultivation. This will ensure that the mineral-nitrogen released from the organic matter is available to the plant at the optimal time. Cultivation at other times particularly if the soil is left bare could result in serious environmental damage from leaching and greenhouse gas production.

Method of incorporation

Surface applications of manures will be less effective than direct incorporation into the soil and will result in greater losses, particularly of ammonium-nitrogen and also cause greater public nuisance from odours.

Slurries

Slurries are not widely used as they generally come from systems of animal housing that are not acceptable to the organic movement. But the direct incorporation of such materials is very efficient at placing the major nutrients where they are required by the crop and also preventing excessive losses, particularly of the volatile components. In some countries, e.g. Germany, the time for application is regulated by the authorities.

Effects on soil structure

Conversion to an organic system will result in increased soil organic matter levels, particularly when red clover residues or large amounts of straw are incorporated. However these increases are not great and so large soil organic matter increases in organically cultivated soils are not anticipated, although soil organic nitrogen content is expected to slowly increase. The anticipated decline in soil phosphorous and potassium levels in legume only based systems can be rapidly corrected by additions of farmyard manure and other materials.

Yields can be affected by soil texture, being higher in more open coarse-textured soils than in finer textured ones. This could be due to a greater rooting zone in the coarse-textured soils resulting in increased nitrogen uptake.

Improved soil aggregation resulting from the introduction of organic matter may thus indirectly lead to an increase in yield.

Combinations of organic and inorganic nitrogen

Although not acceptable in the philosophy of organic farming, an approach that uses combinations of organic and inorganic nitrogen inputs would probably provide a system that is more amenable to environmental management giving the highest yields and the lowest environmental impact overall.

Crop N requirements

On soils with high phosphorous and potassium levels, nitrogen is expected to be the growth-limiting factor. It has been suggested that crop growth just after emergence will be nitrogen-limited as the rooting-zone is relatively small (Kandeler *et al.*, 1994). Enough nitrogen will be accessed when root length densities are more than 1 cm cm^{-3} in a rooting zone that exceeds 60 cm. So at this stage crop growth will not be nutrient limited. As full soil cover is reached crop growth may be nitrogen-limited again as demand exceeds availability. This early limitation of growth due to inadequate nitrogen supply will result in reduced leaf expansion and so light interception. Yield will thus be reduced, by up to 30%, compared to traditional cultural methods.

Soil contribution to nitrogen, phosphorous, potassium, etc.

The supply of such elements to any crop is dependent on soil type (Mackie-Dawson *et al.*, 1990) and soil organic matter content, previous crop residues and microbial activity. All these will need assessing on an individual field basis, but could be expected to increase as the soil fertility increases with the continued incorporation of the plant residues and applied organic matter.

Other logistic considerations

Pests and diseases

The fungal diseases *Phytophthora infestans* and *Rhizoctonia spp.* can cause significant reductions in yields. There may also be late season aphid infestations and problems with the potato beetle *Leptinotarsa decemlineata*. The use of disease resistant varieties that have been pre-sprouted and planted early will reduce these problems (Stockdale *et al.*, 1992). Weed control can be an economic problem in organic farming, but control is possible by mechanical ridging up before canopy closure.

The nematode problem can be minimised by potato cycles of 4 years or more.

The effect of existing crop residues, and the use of green manures?

Restrictions on nitrate levels in ground waters means that the timing of cultivation and the application of manures and residues is critical. Particularly when crops such as potatoes which only 'occupy' the soil for relatively short periods are grown. Organic farming often leads to more post harvest leaching of nitrogen per hectare as mineralisation of manure and other sources of nitrogen increases during the growing season. In contrast chemical fertilisers are mainly supplied at the beginning of the season.

In some situations the use of 'green manures' as sinks for plant uptake of the nitrate in the soil-solution will be necessary.

Microbe/plant interactions

Microbially mediated nitrogen transformations in soils are the major biological source of N_2O. N_2O is a particularly effective greenhouse gas that is also involved in ozone-formation in the troposphere and ozone-destruction in the stratosphere. Microbial activity in soils is normally limited by carbon availability. Organic amendments to soils should reduce such limitations and the enhanced microbial activity may result in increased N_2O production.

The system must be managed to ensure that 'mobilisation' of the nitrogen, from whatever organic source, occurs at the correct time for uptake by the plant. This will ensure that the system is not fuelled into other negative environmental side effects that are also associated with microbial activity in the biogeochemical transformations of nitrogen. These are the same as those found in conventional systems, acidification, eutrophication, ammonia volatilisation, leaching of both nitrogen and phosphorous, heavy metal release to soil, etc.

Rates of nitrogen mineralisation, nitrification, denitrification and the function of other enzymes involved in the nitrogen cycle will be significantly enhanced in organically manured soils (Laanbroek & Gerards, 1991; Wheatley *et al.*, 1997). Nitrification is stimulated in organic farming as a result of an increased supply of ammonium-nitrogen by enhanced mineralisation and also by enhanced heterotrophic activity (Wheatley *et al.*, 1997). Denitrification losses tend to be slightly increased in organically manured systems, denitrification being a carbon dependent process, with losses of between 2% from spring and 12% from winter applications (Oleson *et al.*, 1997).

Regulatory bodies within the EU

Organic crop production across the EU is regulated by EU Regulation 2092/91. Under this each member state has a national authority which then implements the national regulations. In the UK this is done by UKROFS, in The Netherlands by SKAL and in Germany by separate authorities in each state. All these have established protocols that regulate the certification bodies that do farm inspections and certifications

World standards are set by IFOAM. They set minimum standards for trade between countries outwith the EU.

Conclusions

Sustainability needs to be both financial and biological. Organic production of potatoes will only be adopted on a significant scale when it attains at least the profitability of conventional systems. Present profitability is achieved by a combination of price premiums and aid, the future of both is unpredictable.

Timing is the critical factor in all management decisions associated with organic cultivation.

To optimise potato production and nitrogen management in organic systems, so obviating chemical fertilisation it will be necessary to:-

1. Ensure that nitrogen, and the other nutrients, are present in the soil solution at the correct time for plant uptake.

2. Apply organic materials at the correct time of year with rapid incorporation into the soil.
3. Analyse the organic matter before application, no guessing!
4. Use the correct crop rotation.
5. Exploit the benefits of high soil organic matter contents for potato production.

7 Recommendations and trends in research and practice of application of nitrogen and water to potato crops

A.J. Haverkort & D.K.L. MacKerron

This chapter very briefly summarises the role and importance of nitrogen and water in potato production. Then the main conclusions of the various chapters are highlighted. Finally some important trends are described and likely future developments in decision support are discussed. What is the role of computer modelling? The consumers are the end users in a complete agricultural production chain that starts with a production site and a variety through crop management, harvest, storage and processing. Nitrogen and water management exert an influence in each link of this chain. What are the known new sensing techniques and future ones? Information and communication technology is bound to affect nitrogen and water management through decision support systems, but how? Breeding nowadays is assisted by genetic modification. Are drastic changes to be expected that affect the utilisation of water and nitrogen? It is not possible to foretell the future but this chapter serves as a conclusion of the book and hints at research that is needed to achieve continuous advancement in potato production.

Nitrogen and water

The atmosphere is roughly eighty percent gaseous nitrogen. The other principal components are oxygen, water and carbon dioxide. Much is known about the role of water and nitrogen in crop production. Nitrogen in the air is fixed into nitrate or ammonia which can be taken up by plants where nitrogen is essential in the form of amino acids and proteins. Fixation takes place in the atmosphere by lightning and in the soil by bacteria such as *Rhizobium* that lives in conjugation with plants such as clover and lupin or by free living bacteria like *Nitrobacter*. Another important source of nitrogen for living plants is nitrogen from decomposing, dead organic matter. This and synthetically produced nitrate and ammonia are the most important sources of nitrogen in agriculture.

About 3.5% of the dry weight of a potato leaf is nitrogen as protein, mainly Rubisco. Young leaves may contain up to 2% of free nitrate but this usually disappears within a few weeks of the last application of fertilizer nitrogen. The total amount of nitrogen that has to be applied to the crop depends on the amount that the crop will need minus what it will take up from organic sources. The amount and timing of nitrogen fertilisation also strongly influence the quality of the tubers. High amounts of nitrogen lead to low dry matter concentrations which may impair storage and processing behaviour. When too much nitrogen is applied it adversely affects the profitability of the crop and the environmentally friendliness of its production as a sustainable resource. This book has shown the processes that are involved in the uptake of nitrogen and the development of the crop, how they can be measured and calculated in balanced equations, and how all the knowledge acquired to date could be used in decision support systems to aid the farmer, and in legislation to set reasonable limits to the pressure on the environment.

The potato plant is about 75% water which is primarily a solvent and a means of transport for the nutrients that are taken up by the roots. As water evaporates from the leaves, nitrogen, potassium and phosphorus are transported in to the plant. Transpiration also helps to avoid overheating and so reduces the respiration rate and prevents photoinhibition. When water is in short supply, the leaves close their stomata to limit its loss but thereby they also reduce the uptake of carbon dioxide and the efficiency with which light is used. Increased drought will subsequently lead to wilting and reduced exposure of the leaves to the sun. A plant produces about 8 grams of dry matter per litre of water taken up. When water is short this so called water use efficiency increases but total growth slows down. Much research has been carried out to increase the water use efficiency without reducing growth rates but to little avail. Water supply, its amount and timing, strongly influence the quality of the produce. Lack of water around tuber initiation reduces the number of tubers. During tuber bulking it increases the dry matter concentration. Fluctuating water supply during tuber growth often leads to defects such as secondary growth, knobbiness and cracks.

Depending on when and where the crop is grown, the water for potato crops comes in variable proportions from irrigation, natural rainfall and capillary rise from the soil. Losses occur due to evaporation directly from the soil or from the plants, and in drainage. Worldwide, water is an important resource that is becoming scarcer. Conflicts of demand are becoming more frequent and over exploitation of surface and ground water to irrigate crops and for other purposes, adversely affects the surrounding environment. This book gives a brief review of the most important process, how they are measured and quantified in equations and balances, what are the modes of application, and how farmers can be aided by decision support systems.

Recommendations

Strategic decisions

Synthetically produced nitrogen fertiliser only became available in the second half of the twentieth century. This made it possible for farmers to match the demand and supply of the crop more precisely than before. Previously, organic manure was the main source of nitrogen for the potato crop and in some areas it has remained important. We believe that it will become generally more important again. The management of soil organic matter, generally to maintain or increase its levels, is crucial for a number of reasons.. Soil organic matter improves the physical condition of the soil, increasing aggregate stability, improving the water holding capacity, and conserving the soil. It also improves the chemical soil fertility as it can provide a steady supply of plant nutrients. The soil organic matter is continually being oxidised and would gradually disappear if new organic matter were not added to provide a turn-over. Sources of fresh organic matter (apart from off-farm organic waste from cities and industries) are slurry and farm yard manure, harvest residues and green manures. Strategic decisions are not related to a specific year nor a specific field but depend on the type of farming operation and on the type of farmer. The best strategic management and planning to make optimal use of soil organic matter depends on the type of farm (arable, mixed, integrated, organic) its size and the possibilities to vary the rotation. The availability of organic matter from off the farm should also be considered. The following considerations apply:

- If the soil organic matter content is low and none is available on the farm then organic matter should be imported.
- The incorporation of crop residues (potato haulm and cereal straw) is encouraged.
- Reduction in the proportion of crops, such as potato and sugar beet, that leave little organic residue needs to be considered.
- As the potato crop takes up nitrogen most rapidly at the beginning of its growth cycle and as mineralization continues thereafter, it may leave more

nitrate than some other crops. The use of catch crops to follow potato should reduce the leaching of nitrogen. They have the added benefit of improving the soil structure.

- In organic farming systems, clover or other leguminous crops need to be included in the rotation to avoid a collateral built-up of phosphorus in the soil that, too, will eventually leach from the soil.
- In future, off farm organic waste from cities and industry will become more important.

Tactical decisions

Tactical decisions call for information in the year potatoes are grown. What a farmer needs to know before actually applying nitrogen to a particular field in a specific year are:

- What will be the demand of the crop ? A late crop needs a longer growing season, generally a higher yield, and has a higher nitrogen demand than early or short-season crops. The potential yield also depends on the availability of irrigation. Early and short-season crops may be grown from early varieties or may be killed prematurely, as is the case with seed potatoes or when the market calls for 'new potatoes'.
- How much nitrogen is present in the soil prior to planting and before any fertilizer nitrogen is applied? Soil analysis at the end of winter is the best way to find this out.
- What is the expected rate of mineralization from soil organic matter? This depends on the type of organic matter previously incorporated in the soil.
- What are the expected losses (mainly due to leaching). These depend on patterns of rainfall and irrigation.
- A decision has to be taken whether to apply all the nitrogen fertilizer at planting or to apply it in split doses. We recommend the use of split applications but then the crop should be monitored, and maybe the soil, so as to fine tune the second or later dressings.
- Tactical decisions on when and how to apply nutrients are critical in avoiding losses of nitrogen. These include how to manage the water supply to the crop and when and how to sow catch crops to make use of the nitrogen in the next growing season.

Operational decision support

Once any organic manure has been applied followed by an initial nitrogen dressing and the potatoes have been planted, only operational decisions on nitrogen and water supply remain to be taken by the farmer. Support for his decisions comes from an array of monitoring and sampling techniques for the soil and the crop. Invasive methods include the destructive sampling of plant parts to determine their nitrate and total nitrogen concentrations. Recently, other non-invasive techniques have been developed from which measurements correlate well with nitrogen concentration and, even, uptake. Such techniques help farmers to determine whether or not the crop needs additional fertilisation. Soil sampling helps to determine whether the soil nitrate pool is nearing depletion. Supplemental dressings should be given taking these measurements into account. The dressings are only useful when the additional nitrogen reaches the plant in time to extend its growth. Foliar dressings are helpful, being rapidly, but incompletely absorbed. Soil dressings require rainfall or irrigation shortly after application. They are taken up less quickly and, again, will not all be absorbed by the crop. The combined application of water and nitrogen through fertigation is currently being studied in several countries.

Future developments in decision support

Sensing and models

There is an increased need for decision support systems for a number of reasons. With application of water and nitrogen being subject to restriction, it is important to make the fullest use of available knowledge in order to make the best use of less or fewer resources. A second reason is that as farms become larger the farmers benefit more from assistance with their decisions. These decisions are valid for increasingly larger areas and so cost less per unit area. Thirdly, and maybe most importantly, improved access to knowledge as such and to knowledge aggregated in models, and the availability of sensing techniques and of information and communication technology can improve the services given by extension agents. Lastly, governments are rapidly reducing their direct support of the primary production sector of society, especially the farming community. Many private consultancy firms are being formed and they will serve farmers best if tools are available for decision support. These four reasons are likely to become more important in the coming years

Quantitative information is needed to estimate a crop's requirement for nitrogen and water. Some of this information is generic and used for strategic planning such as when to plant and to harvest as these are determined by the long term weather data. Similarly, soil type can influence the effects of increased application of organic manures over the long-term. Time and site specific information is used for operational and tactical crop management decisions. The soil organic matter, and depth of the profile determine the soil's water holding capacity. Sampling soil nitrogen pre-planting is a common basis for the decision on how much nitrogen should be applied pre-planting. Soil sampling is, again, important together with measurements of crop nitrogen (or an analogue such as crop reflectance) when assessing the need for supplemental dressings of nitrogen. Water use may be monitored by field specific water balances or by modern sensing techniques such as small weather stations, electronic sensors of soil water, and leaf thermometry.

Models are abstractions of reality, reductions, and yet a lot of quantitative information is needed to run them. Fortunately, most of it is generic information, not unique to that crop, such as weather and soil parameters. Rate variables in any model that is likely to be used in decision support systems include the conversion rate of solar energy into dry matter and the water use efficiency. Both of these depend on the availability of water, solar radiation, and nitrogen. Many of the quantitative interactions between these variables are known or are rapidly being included in current models. For decision support, e.g. to decide how much water or nitrogen to apply, models need more inputs than just generic data. Data are needed on planting and emergence and the progress simulated by a model has to be confirmed by observation to give confidence in the predictions of water or nitrogen uptake and of yield. This is where sensing techniques come into their own; techniques such as infrared reflectometry for proportional ground cover, electronic probes for soil water content, and ratios of reflection of various wavelengths of light to determine the nitrogen concentration of the standing biomass. Calculation of response to a certain amount of nitrogen or water about to be applied will require that a model that previously ran on current data should switch to some estimate of likely future weather such as long term averages. Such a procedure would also allow the calculation of risks by calculating consequences of treatments for years that are drier or wetter than average.

Information and communication technology

The complete chain of agricultural production and the role of modelling in it is depicted in Figure 1. It has the consumer, shown at the top of the figure, as the most important feature. It is the consumer who drives the need to process potatoes into desired products.

Modelling agro-production chains

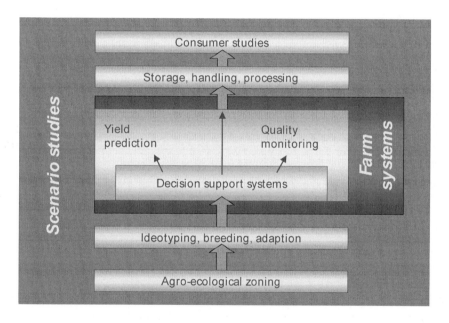

Figure 1   The role of modelling in the potato production chain.

The fresh or processed potato on the shelf arrives there after production and a storage period. The requirements of the consumer determine the prior storage, handling and processing. These, in turn, determine the manner of production, its yield and quality. The several market outlets, starch industry, deep frozen French fries industry, and fresh table potato each have their own requirements for quality and their own preferred varieties. A chain starts with a place and a time of production. The high priced, earliest potatoes are produced in winter and spring in southern Europe. The second earliest potatoes come from the Atlantic coast where frosts are less severe and end sooner, yet the growing season is kept short so that the tubers can reach the market while prices are high.. Processing potatoes require a longer growing season to achieve the higher dry matter concentrations required for processing. Starch potatoes, too, are grown to produce the maximum dry matter, in the form of starch. As already explained, the management of the water and nitrogen supplies differ for the different destinations of the tubers. The production chain, therefore, dictates the water and nitrogen application regimes and modelling can have a place in indicating which regimes are appropriate for each market outlet.

Increasing amounts of data are collected in the course of crop production. Some of these are site and time specific and are produced as a consequence of decisions such as nitrogen and water requirements within the course of a growing season. Traceability in production chains also requires an increased amount of data to be registered. Similarly, the requirement to meet certain standards for product registration and certification, as in organic farming, demands additional record keeping.

There is, now, an increased desire to capitalise on these data and to turn them into a self learning system. Nowadays many of the data are stored in digital form. Extension firms serving farmers with decision support systems keep most information on levels of nitrogen and water in soil and crops available in automated computer based systems. Records of yield and quality (dry matter and nitrogen concentration, frying colour, protein content,..) are similarly kept by the processing industries. Such field-specific information would be ideally suited to a self learning system that gets better after each year in which data are collected (Fig. 2) as the decisions may be tailored to an individual site. Figure 2 shows the requirements and working of such a system. Data are collected annually on a field and stored in a database. With the aid of quantitative approaches (statistical and mechanistic models) the data are interpreted using yield-gap analysis and other techniques and suggestions are generated for management of water or nitrogen. Before being implemented, this advice is tested in scenarios with long-term average weather data for robustness and risk of failure (over or under irrigation and over or under nitrogen fertilisation). Once the recommendation has been implemented the response is recorded for subsequent years and the database, and subsequent recommendations, are adjusted accordingly.

Self-learning systems

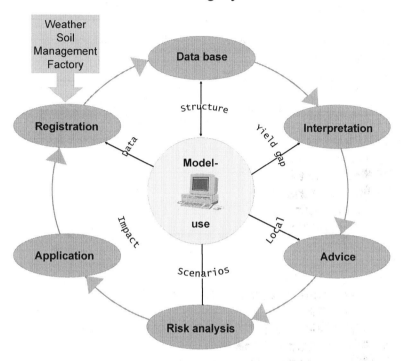

Figure 2   Schematic representation of a self learning or self adjusting sytem making use of automated generation and transfer of knowledge.

An advantage of the whole system is that a continuum is created between farmers, extension agencies collecting and interpreting the data, and science because all three parties exchange data and views and learn and improve while contributing to the system.

References

AACC (American Association of Cereal Chemists), 1995. Approved method 46-30. AACC, St. Paul, USA.

Addiscott TM & Powlson DS, 1992. Journal of Agricultural Science, Cambridge 118: 101-107.

Allison FE, 1973. Soil organic matter and its role in crop production. Elsevier Sc. Publ. 637 pp.

Ampe G, 1991. Mededeling 324. P.O.V.L.T., Roeselare, Belgium, 4 pp.

Anne P, 1945. *Annales Agronomiques* 15: 161-172

Anonymous, 1980. Vlugschrift voor de landbouw 317:1-12

Anonymous, 1989. Handboek voor de Akkerbouw en de Groenteteelt in de Vollegrond. PAGV, Lelystad, The Netherlands, 252 pp.

Anonymous, 1991. Official Journal of the EC, NO. L375, 31/12/1991, 0001-0008.

Anonymous, 1994. MAFF Reference book 209, H.M. Stationery Office, London, UK, 112 pp.

Anonymous, 1998. In: NMI Praktijkgids bemesting, Wageningen, The Netherlands, 1-60.

Appel T, Siak I & Hermanns-Sellen M, 1995. *Plant and Soil* 176 197-203.

Aulakh MS, Doran JW & Mosier AR, 1992. *Advances in Soil Science,* Volume 18, 57 p.

Bailey J, 1987. In Carr, M.K.V. & Hamer, P.J.C. (eds.) Irrigating Potatoes. UK. Irrigation Technical Monograph 2. pp.61-65.

Bailey RJ, 1990. Irrigated crops and their management. Farming Press, Ipswich, UK, 192 pp.

Bailey RJ, Lord EI & Williams D, 1992. *Aspects of Applied Biology* 33 9-14

Baumgärtel G, 1997. *Kartoffelbau*, 48(1/2), 30-33.

Beare MH, Coleman DC, Crossley DA, Hendix PF & Odum EP, 1995. In: The significance and regulation of soil biodiversity. Kluwer Academic Publisher pp. 5-22.

Ben-Asher J, Meek DW, Hutmacher RB & Phene CJ, 1989. *Agronomy Journal* 776 -781

Biemond H & Vos J, 1992. *Annals of Botany*, 70, 37-45.

Bodlaender KBA, Lugt C & Marinus J, 1964. *European Potato Journal* 7: 57-71.

Boon R, 1981. *Pedologie*, 21, 347-363.

Bradstreet RB, 1965. The Kjeldahl Method for Organic Nitrogen. Academic Press, London.

Breimer T, 1989. Stikstofbijmestsysteem (NBS) voor enige vollegronds-groentegewassen. IKC-AGV, Lelystad, The Netherlands, 58 pp.

Bries J, Vandendriessche H & Geypens M, 1995. Bemesting en beregening van aardappelen in functie van opbrengst en kwaliteit. IWONL, Brussels, Belgium, 250 pp.

Bundy LG, Wolkowski RP, Weis GG, 1986. *American Potato Journal* 63: 385-397.

Büning-Pfaue H, Hartmann R, Harder J, & Urban C, 1998. *Fresenius' Journal of Analytical Chemistry* 360: 832-835.

Burton WG, 1989. The Potato. Longman Scientific & Technical Press. p 204.

Campbell R, 1998. In *Proceedings of the 4th International Conference on Precision Agriculture*. Eds. PC Robert, RH Hurst & WE Larson. Madison, WI, USA: American Society of Agronomy (*in press*).

Campbell GS, 1985. In: Marshall, B and Woodward, Fl.(eds). Instrumentation for environmental physiology. Society for Experimental Biology. Seminar Series No: 22, pp 193-214.

Carter JN, Jensen ME & Bosma SM, 1974. Agronomy Journal, 66: 319-323.

CEN (European Committee for Standardisation), 1997. European Standard EN Brussels Belgium, pp. 12014.

Cerovic ZG, Goulas Y, Gorbunov M, Briantais JM, Camenen L & Moya I, 1996. *Remote Sensing of Environment* 58: 311-321.

Chambers BJ, Lord EI, Nicholson FA & Smith KA, 1999. Soil Use and Management 15: 137-143

Clevers JGPW & Van Leeuwen HJC, 1996. *Remote Sensing of Environment* 56: 42-51.

Cormack WF, 1997. Proceedings 3rd ENOF Workshop, Ancona, Italy.

Curwen D, 1993. Water Management. in Rowe & Randall C (eds.) Potato Health Management. The American Phytopathological Society, pp 70-71.

Dalla Costa L, Delle Vedove G, Gianquinto G, Giovanardi R & Peressotti A, 1997. *Potato Research* 40: 19-34.

Dampney PMR, Goodlass GG, Riding AE & Froment MA, 1997. *Proceedings 11th World Fertiliser Congress 'Fertilization for sustainable plant production and soil fertility'*, Ghent, September 1997.

De Neve S, Pannier J & Hofman G, 1994. *European Journal of Agronomy* 3: 267-272.

Demeyer P, 1993. PhD thesis, Faculteit Landbouwkundige en Toegepaste Biologische Wetenschappen, Gent, Belgium, 235 pp.

Denmead OT, 1983. In: Freney JR & Simpson JR (eds.). *Developments in plant and soil science* 9: 133-157.

Doorenbos J & Kassam AH, 1979. Yield response to water. *FAO Irrigation and Drainage Paper* 33.

Dosch P & Gutser R, 1996. *Fertilizer Research* 43: 165-171.

Dyson PW & Watson DJ, 1971. *Annals of Applied Biology* 69: 47-63.

Ecetoc, 1994. Ammonia emissions to air in Western Europe. Technical report No. 62, European Centre for Ecotoxicology and Toxicology of Chemicals, 195 pp.

EEA/UNEP, 1997. Water stress in Europe. EEA Denmark

EEC, 1991. Council Directive concerning the protection of waters against pollution caused by nitrates from agricultural sources (91/676/EEC; Nitrate Directive).

EU, 1998. Freshwater: A Challenge for Research and Initiative EUR18098 EN

Foley MF, 1987. In Carr, M.K.V. & Hamer, P.J.C. (eds.) Irrigating Potatoes. UK. Irrigation Technical Monograph 2. pp.49-53.

Gandar PV & Tanner CB, 1976. *Crop Science* 16: 534-538.

Goffart JP & Guiot J, 1996. In: *Abstracts of Conference Papers*, 13th Triennial Conference of the European Association for Potato Research, Veldhoven, The Netherlands, 14-19 June 1996, 391-392.

Goffart JP, Ninane V & Guiot J, 1992. In: François E, Pithan K & Barthiaux-Thill N (eds.). Nitrogen cycling in cool and wet regions of Europe. Workshop cost, October 1992, Gembloux, Belgium, pp. 132-133

Goffart JP, Olivier M & Destain JP, 1999. In: *Abstracts of Conference Papers*, 14th Triennial Conference of the European Association for Potato Research, Sorrento, Italy, 2-7 may 1999.

Goffart JP, Rouxhet F & Guiot J, 1997. In: Van Cleemput O, Haneklaus S, Hofman G, Schnug E & Vermoesen A (eds.) *Proceedings 11th International World Fertilizer Congress of CIEC, Volume I*, September 7-13, 1997, Gent, Belgium, 669-678.

Goodlass GG & Johnson PA, 1997. *11th World Fertiliser Congress 'Fertilization for sustainable plant production and soil fertility'*, Ghent, September 1997.

Gregory PJ & Simmonds LP, 1991. Water relations and growth of potatoes. In *The Potato Crop*. Ed. Harris, P.M., Chapman & Hall, London, pp. 214-244

Harris PM, 1978. The potato crop. Chapman & Hall, London, 620 pp.

Hassink J, 1992. *Biology and Fertility of Soils* 14: 126-134.

Hébert J, 1973. *Science du Sol* 3: 199-206.

Hillhorst M & Dirksen C, 1994. Dielectric water content sensors: time domain versus frequency domain. In: Prime Demain Reflectometry in Environmental, Infrastructure and Mining Applications v.s. Bureau of Mines, Special Publication SP 19-49, pp. 23-33.

Hofman G, 1977. In: De Leenheer L et ses collaborateurs (eds.) Structure et fertilité des sols limoneux sur fermes mécanisées (étude faite de 1961-1975). Faculty of Agricultural Sciences, Gent, Belgium, 255-290.

Hofman G, 1983. PhD thesis, Faculteit van de Landbouwwetenschappen, Gent, Belgium, 183 pp.

Hofman G, Verstegen P, Demyttenaere P, Van Meirvenne M, Delanote P & Ampe G, 1993. In: Fragoso MAC & Van Beusichem ML (eds.) Optimization of Plant Nutrition. Kluwer Academic Publishers, Dordrecht, The Netherlands, 359-365.

Hooker WJ, 1983. Compendium of Potato Diseases. The American Phytopathological Society pp 12-13

ICC, 1994. Standard No. 105/2. ICC (International Association for Cereal Chemistry), Vienna, Austria.

James DW, Hurst RL, Westermann DT & Tindall TA, 1994. *American Potato Journal* 71: 249-265.

Jarvis SC, Stockdale EA, Shepherd MA & Powlson DS, 1996. *Advances in Agronomy* 57: 187-235.

Jefferies RA, 1989. *Journal of Experimental Botany* 40, 1375 – 1381.

Jefferies RA, 1993. *Agricultural Systems* 41, 93 - 104.

Jefferies RA, 1993. *Annals of Applied Biology* 122, 93 - 104.

Jenkins PD & Nelson DG, 1992. *Potato-Research* 35:127-132.

Jenkinson DS, 1990. *Soil Use and Management* 6: 56-61.

Jensen C, Stougaard B & Olsen P, 1994. *Acta Agriculturae Scandinavica* 44: 75-83.

Johnson PA, Shepherd MA & Lord EI, 1996. *Abstracts conference papers EAPR* Veldhoven, 1996.

Jones JP & Painter CG, 1974. University of Idaho Co-operative Extension Service, Bingham County, Idaho, pp 1-4.

Kandeler E, Eder G & Sobotik M, 1994. *Biology & Fertility of Soils* 18: 7-12

Kay M, 1986. Surface Irrigation systems and practice. Cranfield Press, Cranfield Bedford, 142 pp

Kay M, 1988. Sprinkler Irrigation equipment and practice. B T Batsford Limited, London, 120 pp

Keulen H van & Penning de Vries FWT, 1993. International Crop Science I (eds. D.R.Buxton et al.). Crop Science Society of America, Inc. Madison, Wisconsin, USA. pp. 139-144.

Kirsten WJ & Gunnar UH, 1983. *Microchemical Journal* 28:529-547.

Kolbe H & Stephan-Beckmann S, 1997. Potato Research, 40, 111-129.

Kolbe H & Stephan-Beckmann S, 1997. Potato Research, 40, 135 - 153.

Kolbe H, Zang WL & Ballüer L,1990. Plant and Soil 124: 309-313.

Kölsch E & Stöppler H, 1990. KTBL-Paper No. 147, Darmstadt, Germany, 120 pp.

Kölsch E, 1996. Report of the 18th Potato Convention, Detmold Germany, pp.14-22.

Laanbroek HJ & Gerards S, 1991. *Biology & Fertility & Soils* 12: 147-153

Lampkin N, 1992. Organic Farming. Farming Press Books.

Lemaire G (Ed.), 1997. Diagnosis of the nitrogen status in crops. Springer-Verlag, Berlin, Heidelberg, 1997. 239 pp.

Lesczynski & Tanner, 1976. *American Potato Journal* 53, 69 – 78.

Levy D, 1992. *Annals of Applied Biology* 120, 547 – 555.

Lewis RJ & Love SL, 1994. *HortScience* 29: 175-179.

Loon CD van & Houwing JF, 1989. PAGV-publikatie. Lelystad, PAGV 42:1-90.

Lord EI, Johnson PA & Archer JR, 1999. The Pilot Nitrate Sensitive Areas: Final Report. MAFF, London.

Lorenz H-P, Schlaghecken J & Engl G, 1985. *Deutscher Gartenbau* 13: 646-648.

MacKerron DKL & Peng ZY, 1989. In: *Aspects of Applied Biology 22, Roots and the Soil Environment*. Association of Applied Biologists, Wellesbourne. pp. 199-206.

MacKerron DKL, 1987. In Carr, M.K.V. & Hamer, P.J.C. (eds.) Irrigating Potatoes. UK. Irrigation Technical Monograph 2. pp. 54-60.

MacKerron DKL, Young MW & Davies HV, 1993. *Plant and Soil* 155: 139-144.

MacKerron DKL, Young MW & Davies HV, 1995. Plant and Soil 172: 247-260.

Mackie-Dawson LA, Millard P & Robinson D, 1990. *Plant & Soil* 125: 159-168.

MAFF, 1997. Survey of Irrigation of Outdoor Crops in 1995 (amended) England *Statistics 222/96*

Marshall B, Crawford, JW & McNicol, JW, 1994. In: *Scottish Crop Research Institute, Annual Report 1993*, edited by D. A. Perry and T. D. Heilbronn, Invergowrie, Dundee: Scottish Crop Research Institute. p. 63-65.

Marshall B, Crawford, JW & McNicol, JW. 1995. In: *Potato ecology and modelling of crops under conditions limiting growth*, edited by A. J. Haverkort and D. K. L. MacKerron, Netherlands. :Kluwer Academic Publishers, p. 323-340.

McDonald AJ, Poulton PR, Powlson DS & Jenkinson DS, 1997. *Journal of Agricultural Science, Cambridge* 129: 125-154.

McTaggart IP & Smith KA, 1996. Transactions of the 9th Nitrogen Workshop held in Braunschweig, September 1996, 523-526.

Millard P & MacKerron DKL, 1986. *Annals of Applied Biology* 109: 427-437.

Millard P & Marshall B, 1986. *Journal of Agricultural Science, Cambridge* 107: 421-429.

Miller DE & Martin MW, 1987. *American Potato Journal* 64: 109-117.

Mitchell HL, 1972. *Journal of the Association of Official Analytical Chemists* 55: 1-3.

Monteith JL, 1965. In: *State and movement of water in living organisms*, Anonymous XIXth Symposium of the Society of Experimental Biology, pp. 205 - 234.

Müller K, 1977. *Kartoffelbau* 28: 4-6

Neeteson JJ & Wadman WP, 1987. *Fertilizer Research* 12: 37-52.

Neeteson JJ & Wadman WP, 1991. Nota 237, Instituut voor Bodemvruchtbaar-heid, Haren, The Netherlands, 18 pp.

Neeteson JJ & Zwetsloot HJC, 1989. *Netherlands Journal of Agricultural Science* 37: 129-141.

Neeteson JJ, 1989. *Netherlands Journal of Agricultural Science* 37: 143-155.

Neeteson JJ, 1989. *Netherlands Journal of Agricultural Science* 37: 227-236.

Neeteson JJ, 1992. In: Adriano DC (ed.) *Advances in Environmental Science*, Groundwater Series, Springer, New York.

Neeteson JJ, 1995. In: Bacon PE (ed.) Nitrogen fertilization in the environment. Marcel Dekker Inc., New York, USA, pp. 295-325.

Neeteson JJ, De Wijngaert KEL, Belmans CFM & Groot JJR, 1988. Nota 186, Instituut voor Bodemvruchtbaarheid, Haren, The Netherlands,17 pp.

Neeteson JJ, Greenwood DJ & Draycott A, 1989. *Netherlands Journal of Agricultural Science* 37 237-256

Neeteson JJ, Wijnen G & Zandt P, 1984. *Bedrijfsontwikkeling*, 15: 331-333.

Nelson DW & Sommers LE, 1982. In: Page AL, Miller RH & Keeney DR (eds.) Methods of Soil Analysis. Part 2 Chemical and Microbiological Properties. American Society of Agronomy, Madison, Wisconsin, USA, 539-580.

Neubauer W, 1997. In: Pötke E & P Schuhmann (Eds.): Fresh Potatoes. AgriMedia, Holm, Germany, pp. 52-54

Ninane V, Goffart JP, Meeus K, Guiot J & François E, 1995. In: Geypens M & Honnay JP (eds.) Landbouwkundige en milieugerichte functies van de organische stof in de bodem. IWONL, Brussels, Belgium, pp. 67-104.

Nitsch A & Varis E, 1991. *Potato Research* 34: 95-105.

Norman JM & Jarvis PG, 1975. *Journal of Applied Ecology*, 12: 879-891.

O'Sullivan J & Reyes AA, 1980. *Journal of American Society for Horticultural Science* 105: 809-812.

Oleson T, Griffiths BS, Henriksen K, Moldrup P & Wheatley RE, 1997. *Soil Science* 162: 157-168

Oparka KJ & Wright KM, 1988. *Planta* 174: 123 – 126.

Osaki M, Yoshida M, Nakamura T & Tadano T, 1993. *Soil Science and Plant Nutrition* 39: 595-603.

Osborne BG & Fearn T, 1986. Near infrared spectroscopy in food analysis. Longman Scientific & Technical, New York.

Paltineanu IC & Starr JL, 1997. *Soil Science Society of America Journal* 61: 1576-1585.

Pannier J, Hofman G & Vanparys L, 1996. In: Van Cleemput O, Hofman G & Vermoesen A (eds.) Progress in Nitrogen Cycling Studies. Kluwer Academic Publishers, Dordrecht, The Netherlands, pp. 353-358.

Penman HL, 1948. *Proceedings of the Royal Society, Series B* 193 A:120 - 145.

Porter GA & Sisson JA, 1991. *American Potato Journal* 68: 493-505.

Porter GA & Sisson JA, 1993. *American Potato Journal* 70: 101-116.

Postma R & Loon CD van, 1996. Transactions of the 9th Nitrogen Workshop held in Braunschweig, September 1996, pp. 535-538.

Powlson DS, 1997. Proceedings of the Fertiliser Society no. 402, 42 p.

Rahn CR, Greenwood DJ & Draycott A, 1996. In 'Progress in nitrogen cycling studies' Ed.van Cleemput et al, Kluwer Academic Publishers 255-258.

Rawlins SL, 1996. In *Precision Agriculture, Proceedings of the 3rd International Conference*, June 23-26, 1996, Minneapolis, Minnesota, PC Robert, RH Hurst & WE Larson (eds.) , pp. 1222.

Ris J, Smilde KW & Wijnen G, 1981. *Fertilizer Research* 2: 21-32.

Roche Diagnostics, 1997. Methods of enzymatic bioanalysis and food analysis. Mannheim, Germany.

Rolo JAC, 1995. Produção Final Consumo Intermediário e Valor Acrescentado Bruto por Agricultura e da Silvicultura em 1990. INIA, DEESA, Oeiras, Portugal, 55 pp.

Rowe RC & Secor GA, 1993. In Rowe, Randall C (ed.) Potato Health
    Management. The American Phytopathological Society, pp. 37.
Salomez J, Hofman G, Delanote P & Ampe G, 1997. In: Martin-Prével P & Baier
    J (eds.) IXth International Colloquium for the Optimization of Plant
    Nutrition, 8-15 September 1996, Prague, Czech Republic, pp. 465-470.
Schleef KH & Kleinhanss W, 1996. Regional nitrogen balances in the EU
    agriculture. Conference Proceedings on the Nitrate Directive and the
    Agricultural Sector in the EU. The Hague, June 20-22.
Schrage R & Scharpf HC, 1987. *Gemüse* 10: 412-414.
Shepherd MA & Lord EI, 1996. *Journal of Agricultural Science, Cambridge* 127:
    215-229.
Shepherd MA, 1992. Proceedings of the Fertiliser Society no. 331: 32 p.
Shepherd MA, Stockdale EA, Jarvis SC & Powlson DS, 1996. *Soil Use and
    Management* 12: 76-85.
Smith KA & Chalmers BJ, 1993. *Soil Use & Management* 9: 105-111
Solomon KH, 1988. Irrigation System Selection, Irrigation Notes, Center for
    Irrigation Technology, California State University, Fresno, USA, 11 pp.
Springer U & Klee J, 1954. *Zeitschrift für Pflanzenernährung, Düngung und
    Bodenkunde* 64: 1-26.
Stafford JV, 1998. In *Protection and production of sugar beet and potatoes.* eds.
    MFB Dale, AH Dewar, SJ Fisher, PPJ Haydock, K Jaggard, MJ May, HG
    Smith, RM Storey & JJJ Wiltshire, *Association of Applied Biologists,
    Aspects of Applied Biology* 52: 79-86.
Stark JC, McCann IR, Westermann DT, Izadu B & Tindall TA, 1993. *American
    Potato Journal* 70: 765-777.
Starr JL & Paltineanu IC, 1998. *Soil Science Society of America Journal* 62:
    114-122.
Stevenson GJ, 1965. In: Black CA (ed.) Methods of Soil Analysis. Part 2
    Chemical and Microbiological Properties. American Society of Agronomy,
    Madison, Wisconsin, USA, 1409-1421.
Stockdale EA, McKinley RG & Rees RM, 1992. *Aspects of Applied Biology* 30:
    387-392
Storey & Davies, 1992. Tuber Quality. In: Harris PM (ed.) The Potato Crop: the
    scientific basis for improvement (2nd edit.). Chapman & Hall. London, pp
    507-569.
Timm H, Bishop JC, Tyler KB, Zahara M, Schweers VH & Guerad JP, 1983.
    American Potato Journal 60: 577-585.
Topp GC, Davis JL & Annan AP 1980. *Water Resources Research* 16: 574-582.

Usher CD & Telling GM, 1975. *Journal of the Science in Food and Agriculture* 45: 1793-1805.

Van Delden A, 1988. AB-DLO Report 82. Research Institute for Agrobiology and Soil Fertility, Wageningen, 55 pp.

Van der Paauw F, 1963. *Plant and Soil* 19: 324-331.

Van Kraalingen DWG, 1997. Report, DLO Research Institute for Agrobiology and Soil Fertility (AB-DLO), The Netherlands, pp. 33

Van Loon CD, 1981. *American Potato Journal* 58: 51-69.

Van Meirvenne M & Hofman G, 1989. *Plant and Soil* 120, 103-110

Van Soest PJ, 1963. *Journal of the Association of Official Analytical Chemists* 46: 829-835.

Vandendriessche H, Geypens M, Bries J & Hendrickx G, 1992. *Landbouw-tijdschrift*, 45: 395-402.

Velthof GL, Erp PJ van & Steevens JCA, 1999. Meststoffen 1999 (In Dutch).

Vilain M, 1996. La production végétale. Vol 2 La maîtrise technique de la production. Lavoisier 449 pp.

Vos J & Bom M, 1993. *Potato Research* 36: 301-308.

Vos J & Groenwold J, 1986. *Plant and Soil* 94:17-33.

Vos J & Groenwold J, 1988. *Annals of Botany* 62: 363 - 371

Vos J & Groenwold J, 1989. *Potato Research,* 32:113-121

Vos J & Oyarzun PJ, 1987. *Photosynthesis Research* 11: 253-264.

Vos J & Putten PEL van der, 1997. *Plant and Soil* 195: 299-309.

Vos J, 1996. *European Journal of Agronomy* 5: 105-114.

Vos J, 1997. *Potato Research* 40: 237-248.

Walkley A & Black IA, 1934. *Soil Science* 37: 29-38.

Walworth JL & Muniz JE, 1993. *American Potato Journal* 70: 579-597.

Ward ER, 1987. In Carr MKV & Hamer PJC (eds.) Irrigating Potatoes. UK. Irrigation Technical Monograph 2. pp. 41-48.

Waterer D, 1997. *Canadian Journal of Plant Science* 77: 273-278.

Weber L, Putz B & Lindhauer MG, 1996. Abstracts of Conference Papers, Posters and Demonstrations; 13th Triennial Conference of the EAPR, Veldhoven, 615-616.

Wheatley RE, Griffiths BS & Ritz K, 1991. *Biology and Fertility of Soils* 11: 157-162.

Wheatley RE, Ritz K & Griffiths BS, 1997. *Biology & Fertility of Soils* 24: 378-383

White RP & Sanderson JB, 1983. *American Potato Journal* 60: 115-126.

Whitehead DC, 1981. *Journal of the Science of Food and Agriculture* 32: 359-
    365.
Williams CMJ & Maier NA, 1990. *Journal of Plant Nutrition* 13: 985-993.
Wischmeier WH & Smith DD, 1978. Predicting Rainfall Erosion Losses: A Guide
    to Conservation Planning. Washington US Department of Agriculture,
    Agriculture handbook, Nr. 537, 58 pp.
Wright JL & Stark JC, 1990. In Stewart B A & Nielsen D R (Eds) Irrigation of
    Agricultural Crops - Agonomy Monograph No 30. ASA-CSSA-SSSA, USA,
    pp. 859-888.
Yiqun Gu, McNicol JW, Peiris DR, Crawford JW, Marshal BI, & Jefferies RA,
    1996. *AI Applications* 10: 13-24.
Young MW, Davies HV & MacKerron DKL, 1993. In: *Proceedings VIIIth
    International Colloquium for the Optimization of Plant Nutrition, Lisbon
    1992.* Kluwer Academic Publishers, Dordrecht, pp. 7-11
Young MW, MacKerron DKL & Davies HV, 1997. *Potato Res.* 40: 215-220.

Addresses of the authors

Carlos Aguilera
University of Almeria
Departamento Biologia Vegetale
E-04071 Almeria
Spain
E-mail: caguiler@ualm.es

Roger Bailey
ADAS Gleadthorpe Research Centre
Meden Vale
Warsop
Nottinghamshire NG20 9PF
United Kingdom
E-mail: roger.bailey@adas.co.uk

Stefano Bona
Dipartimento di Agronomia Ambientale e Produzioni Vegetali
University of Padova
Agripolis – Via Romea 16
35020 Legnaro (Padova)
Italy
E-mal: sbona@agripolis.unipd.it

Remmie Booij
AB Wageningen University and Research Center
P.O. Box 14
6700 AA Wageningen
The Netherlands
E-mail: r.booij@ab.wag-ur.nl

Michel Colauzzi
Instituto di Genetica e Sperimentazione Agraria "N. Strampeli"
Via Marconi N1
36045 Lonigo, Vicenza
Italy
E-mail: colauzzi@agripolis.unipd.it

W.F. Cormack
ADAS
Terrington St Clements
KingsLyn
Norfolk PE34 4PW
United Kingdom

Luisa Dalla Costa
University of Udine
Department of Crop Production
Via delle Scienze 208
33100 Udine
Italy
E-mail: dallacosta@palantir.dputa.uniud.it

Stefaan De Neve
Universiteit Gent
Faculty of Agricultural and Applied Biological Sciences
Department of Soil Management and Soil Care
Coupure 653
B-9000 Gent
Belgium
E-mail: stefaan.deneve@rug.ac.be

Giorgio Gianquinto
Dipartimento di Agronomia Ambientale e Produzioni Vegetali
University of Padova
Agripolis – Via Romea 16
35020 Legnaro (Padova)
Italy
E-mail: gianquin@agripolis.unipd.it

Jean-Pierre Goffart
Centre de Recherches Agronomiques de Gembloux (CRAGx)
Département Productions Végétales
4, rue du Bordia
5030 Gembloux
Belgium
E-mail: goffart@cragx.fgov.be

Giovanni Guarda
Instituto di Genetica e Sperimentazione Agraria "N. Strampeli"
Via Marconi N1
36045 Lonigo, Vicenza
Italy

Norbert U. Haase
Federal Centre for Cereal, Potato and Lipid Research
Schuetzenberg 12
32756 Detmold
Germany
E-mail: potato.bagkf@t-online.de

Anton J. Haverkort
AB Wageningen University and Research Center
P.O. Box 14
6700 AA Wageningen
The Netherlands
E-mail: a.j.haverkort@ab.wag-ur.nl

Georges Hofman
Universiteit Gent
Faculty of Agricultural and Applied Biological Sciences
Department of Soil Management and Soil Care
Coupure 653
B-9000 Gent
Belgium
E-mail: georges.hofman@rug.ac.be

Janusz Igras
Institute of Soil Science and Plant Cultivation
Department of Plant Nutrition and Fertilization
Cartoryskich 8 str.
24-100 Pulawy
Poland
E-mail: ij@iung.pulawy.pl

Paddy Johnson
ADAS
Kirton ARC
24, Willington Rd, Kirton, Boston, Lincs PE20 1EJ
United Kingdom
E-mail: paddy_johnson@adas.co.uk

Tuomo Karvonen
Department of Water Resources Engineering
Helsinki University of Technology
Tietotie Fin-02150
Finland
E-mail: tkarvone@pato.hurt.fi

Jouko Kleemola
Kemira Agro OY
Espoo Research Centre
P.O. Box 44 Fin-02271
Espoo
Finland
E-mail: jouko.kleemola@kemira.com

Donald K.L. MacKerron
Scottish Crop Research Institute (SCRI)
Invergowrie
Dundee DD2 5DA
United Kingdom
E-mail: d.mackerron@scri.sari.ac.uk

Bruce Marshall
Scottish Crop Research Institute
Invergowrie
Dundee DD2 5DA
United Kingdom
E-mail: b.marshall@scri.sari.ac.uk

Fernando Martins
UTAD, Departemento Fitotecnia
Apt. 202
5001 Vila Real Codex
Portugal
E-mail: fmartins@utad.pt

Marguerite Olivier
Centre de Recherches Agronomiques de Gembloux (CRAGx)
Département Productions Végétales
4, rue du Bordia
5030 Gembloux
Belgium
E-mail: olivier@cragx.fgov.be

Romke Postma
Nutrient Management Institute NMI
Department of Soil Science and Plant Nutrition
P.O. Box 8005
6700 EC Wageningen
The Netherlands
E-mail: romke.postma@nmi.benp.wau.nl

Joost Salomez
Universiteit Gent
Faculty of Agricultural and Applied Biological Sciences
Department of Soil Management and Soil Care
Coupure 653
B-9000 Gent
Belgium
E-mail: joost.salomez@rug.ac.be

Mark Shepherd
ADAS Consulting Ltd.
Gleadthorpe Research Centre,
Meder Vale,
Mansfield, Nottinghamshire, NG20 9PF
United Kingdom
E-mail: mark.shepherd@adas.co.uk

Juan Valenzuela
University of Almeria
Departamento Biologia Vegetale
E-04071 Almeria
E-mail: jvalenzu@filabres.ualm.es

Jan Vos
Wageningen University and Research Centre
Department of Crop and Weed Ecology
Haarweg 333
6709 RZ Wageningen
The Netherlands
E-mail: jan.vos@users.agro.wau.nl

Ron E. Wheatley
Soil-Plant Dynamics Unit
Scottish Crop Research Institute
Invergowrie
Dundee DD2 5DA
United Kingdom
E-mail: r.e.wheatley@scri.sari.ac.uk

Håndbog til styring af kvaelstof og vand i kartoffelproduktionen

Resume på Dansk oversat af Poul Erik Laerke

Tildelingen af kvaelstof og vand er nøgleproblemer i kartoffelproduktion. Balancen mellem forsyning og optagelse af kvaelstof og vand i planten kan påvirke afgrødens daekningsbidrag kraftigt eftersom både optimal udbytte og kvalitet afhaenger af korrekt forsyning. Overforsyning og underskud giver problemer for afgrøden. Derudover kan overforsyning belaste miljøet. For vands vedkommende betyder det at en begraenset ressource bliver udnyttet ineffektivt, og der er risiko for udvaskning af naeringsstoffer. For kvaelstofs vedkommende vil der forekomme tab til miljøet. Det er på grund af de mulige skadelige effekter af vand og kvaelstof, at der er og kan komme yderligere regulativer på området. Naervaerende bog er et produkt fra et EU-finansieret samarbejde mellem 18 europaeiske forskningsinstitutioner der skal bidrage til en bedre udnyttelse af vand og kvaelstof i kartoffelproduktionen. Titlen på denne "Concerted Action" er EUROPP - Efficiency in the Use of Resources: Optimisation in Potato Production.

Håndbogen forklarer kvaelstofs og vands rolle i kartoffelplantens vaekst og udvikling i kapitel 1, og de efterfølgende kapitler praesenterer hvilke metoder, der eksisterer for at opnå en korrekt balance mellem input og produktion. Kapitel 2 omhandler kvaelstof i planten (total-N og nitrat-N), hvordan kan indholdet bedst bestemmes med eksisterende destruktive og ikke-destruktive metoder, og kan disse målinger støtte beslutninger, der traeffes om kvaelstofgødskning. Kapitel 3 finder svar på spørgsmål om jordens kvaelstofindhold, både uorganisk og organisk kvaelstof, og fastslår vigtigheden af tilstraekkelig prøveudtagninger. Det forklarer vigtigheden af nitrifikation og denitrifikation. Kapitel 4 beskriver vands rolle i jorden og planten, hvordan vandstatus kan bestemmes i hvert medium, og hvordan disse målinger kan bruges i praksis. Kapitel 5 kaster et tilbageblik over eksisterende beslutningsstøttesystemer for kvaelstof- og vandforsyning, hvilke typer kvaelstof kan anvendes (husdyrgødning og kunstgødning), hvilken vandingspraksis er passende, og hvilke beslutningsstøttesystemer er tilgaengelige. De vigtige forskelle mellem kartoffelproduktioner, der anvender og ikke anvender kunstvanding, bliver også taget i betragtning. Kapitel 6 udforsker mulig fremtidig udvikling: Hvilken betydning får afgrødevaekstmodeller og computersimulering

? Hvad er risici og laeren af den nuvaerende praksis ? Hvad kender vi til konsekvenserne af økologisk kartoffelproduktion ? Hvad er mulighederne i praecisionsjordbrug ? Hvad mangler vi at vide ? Bogen konkluderer på baggrund af den nyeste viden, inklusiv den, der er udsprunget fra EUROPP.

Bogen opdaterer viden på en simpel og forståelig måde og den er primaert beregnet for konsulenter, men kan også vaere udbytterig for kartoffelavlere, forskere, politikere og andre beslutningstagere.

Stikstof- en waterbeheer in de aardappelteelt

Samenvatting in het Nederlands vertaald door Joost Salomez

Water- en stikstoftoediening dragen in belangrijke mate bij tot het welslagen van een aardappelteelt. Het onjuist afstemmen van deze toedieningen op de gewasbehoeften heeft een belangrijke invloed op het rendement van het gewas omdat het opbrengstniveau en de kwaliteit van de knollen in sterke mate afhangen van een nauwkeurige toediening van zowel water als stikstof. Zowel overaanbod als tekorten geven bij dit gewas aanleiding tot problemen. Bovendien leidt een overaanbod tot milieuproblemen. In het geval van water betekent dit dat een schaarse hulpbron niet efficiënt aangewend wordt en aanleiding kan geven tot drainageverliezen. Wordt te veel stikstof aangewend, dan zal dit onvermijdelijk uitgespoeld worden naar het grondwater. Omwille van deze mogelijke ongunstige milieu-effecten kunnen zowel stikstoftoediening als wateronttrekking het voorwerp uitmaken van een regelgeving door overheids-instanties. Dit boek werd opgesteld in de loop van een door de EU gefinancierde Concerted Action, met betrekking tot het efficiënt gebruik van stikstof en water in de aardappelteelt, genaamd EUROPP - Efficiency in Use of Resources: Optimisation in Potato Production.

Hoofdstuk 1 van dit handboek behandelt de essentiële rol van stikstof en water voor de groei en de ontwikkeling van een aardappelgewas, terwijl de overige hoofdstukken een overzicht geven van de procedures die beschikbaar zijn om tot een correcte balans te komen van de optimale verhouding tussen inputs en productie. Hoofdstuk 2 behandelt het verloop van de stikstoftoestand van de plant (totale N zowel als nitraat), hoe de N toestand kan bepaald worden door zowel destructieve als niet-destructieve technieken, hoe de bemonstering best wordt uitgevoerd en hoe de bekomen resultaten dienen als ondersteuning van de N-bemesting. Aansluitend hiermee behandelt Hoofdstuk 3 de stikstoftoe-stand van de bodem, waarbij zowel organische- als minerale N beschouwd worden alsmede het belang van een aangepaste bemonstering. Het belang van nitrificatie en denitrificatie wordt besproken en hoe N-verliezen uit de bodem zich voordoen. Hoofdstuk 4 behandelt de rol van water in de plant en in de bodem, hoe de vochttoestand in beide kan gemeten worden en hoe deze metingen in de praktijk gebruikt kunnen worden. Hoofdstuk 5 geeft een overzicht van de bestaande adviessystemen met betrekking tot stikstof- en watertoediening, welke stikstofbronnen gebruikt kunnen worden (organische zowel als minerale), welke irrigatiestrategieën gevolgd kunnen worden en welke beslissingsondersteunende systemen (DSS) beschikbaar zijn voor stikstof en water. Ook wordt aandacht besteed aan de verschillen tussen geïrrigeerde en niet-geïrrigeerde teelten. Hoofdstuk 6 tenslotte gaat in op mogelijke ontwikkelingen. Welke rol is weggelegd voor gewasgroeimodellen en computersimulaties? Welke risico's zijn er verbonden aan de huidige teeltmethodes en welke lessen kunnen daaruit getrokken worden? Welke implicaties heeft biologische landbouw voor de aardappelteelt? Welke voordelen levert ons precisielandbouw? En waarover moeten we nog nieuwe kennis verwerven? Het boek besluit met perspectieven die recente ontwikkelingen ons bieden, met inbegrip van deze die de Concerted Action naar voor heeft gebracht.

Het boek wil op een overzichtelijke en begrijpelijke manier de huidige kennis rond stikstof- en waterbeheer in de aardappelteelt samenbundelen. Het is dan ook bedoeld als een leidraad voor landbouwers, landbouwvoorlichters, onderzoekers en wetgevers en dit elk op hun eigen terrein.

Vesi- ja typpitalouden säätely perunanviljelyssä

Tiivistelmän Suomentanut Maija Paasonen-Kivekäs

Veden ja typen käytöllä voidaan merkittävästi vaikuttaa perunanviljelyn kannattavuuteen, koska sekä sadon määrä että laatu ovat ratkaisevasti riippuvaisia näistä kasvutekijöistä. Liiallisesta veden tai typen käytöstä saattaa seurata ympäristöongelmia. Runsas kastelu kuluttaa turhaan rajallisia vesivaroja ja aiheuttaa ravinteiden huuhtoumariskin. Liiallinen lannoitus voi myös johtaa typen huuhtoutumiseen pinta- ja pohjavesiin. Näiden mahdollisten haittavaikutusten vuoksi viranomaiset pyrkivät säätelemään sekä typpilannoitusta että kasteluveden ottoa.

Tämä käsikirja tehtiin EU:n rahoittaman yhteistyöhankkeen toimesta. Hankkeen nimi oli EUROPP (Efficiency in the Use of Resources: Optimisation in Potato Production) ja sen tavoitteena oli tehostaa typen ja veden käyttöä perunanviljelyssä. Kirjaan on koottu olemassaolevaa tietoa yksinkertaisessa, helposti ymmärrettävässä muodossa. Se on tarkoitettu viljelijöille, maatalousneuvojille, tutkijoille ja päätöksentekijöille.

Kirjan luku 1 käsittelee veden ja typen merkitystä perunan kehitykselle ja kasvulle. Kirjan muissa luvuissa käydään läpi menettelytapoja, joilla voidaan saavuttaa tasapaino veden ja typen käytön ja sadon ja ympäristön välillä. Luku 2 käsittelee kasvien typpitaloutta (kokonaistyppi ja nitraatti). Siinä käsitellään mm. kasvustonäytteen ottoa sekä menetelmiä, joilla voidaan määrittää kasvien ottama typen määrä. Lisäksi arvioidaan, kuinka näitä mittauksia voidaan käyttää hyväksi typpilannoitusta koskevassa päätöksenteossa. Luvussa 3 esitetään vastaavasti maaperän typpeen liittyviä kysymyksiä kuten epäorgaanisen ja orgaanisen typen lähteitä sekä nitrifikaatiota, denitrifikaatiota ja huuhtoumista. Myös tässä yhteydessä tulee esille edustavan näytteenoton merkitys. Luku 4 käsittelee maan ja kasvien vesitaloutta sekä vesipitoisuuden mittaamista ja mittausten hyödyntämistä käytännössä. Luvussa 5 esitetään typpilannoitusta ja kastelua koskevan päätöksenteon apuvälineitä sekä erilaisten typenlähteiden ja kastelumenetelmien käyttöä. Siinä tarkastellaan myös kastelua käyttävän ja pelkästään sadantaan perustuvan viljelyn keskeisiä eroja. Luku 6 luo katsauksen perunantuotannon mahdollisiin tulevaisuuden haasteisiin: Mikä on satomallien ja tietokonesimuloinnin rooli? Mihin nykyisten viljelymenetelmien käyttö johtaa? Mitä tiedetään luonnonmukaisen viljelyn seurauksista? Mitkä ovat täsmäviljelyn mahdollisuudet? Mitä on tutkittava vielä lisää? Lopuksi esitetään viljelyn uusimpia kehitysnäkymiä, joista osa sai alkunsa em. yhteistyöhankkeessa.

Gestion de l'azote et de l'eau en production de pomme de terre

Resumé en Français traduit par J.P. Goffart

La fourniture en eau et en azote constitue un problème majeur en production de pomme de terre. Il existe un équilibre entre l'approvisionnement en ces deux facteurs et les recettes qu'ils procurent. Cet équilibre influence fortement la rentabilité de la culture car le niveau de rendement et la qualité des tubercules dépendent étroitement de la fourniture correcte en eau et en azote. Une fourniture excessive ou un déficit en ces éléments mèneront à des problèmes culturaux. De plus, des excès induisent des problèmes environnementaux. Dans le cas de l'eau, cela signifie une utilisation inefficiente de cette ressource limitée pouvant induire des pertes par infiltration. Dans le cas où l'apport d'azote est trop élevé, l'excès sera entrainé par lessivage vers la nappe phréatique. L'existence de ces effets négatifs potentiels explique que les apports en azote et l'extraction de l'eau peuvent faire l'objet de réglementation de la part des autorités gouvernementales. Ce livre a été réalisé durant une Action Concertée subsidiée par l'Union Européenne, axée sur l'utilisation efficiente de l'azote et de l'eau en culture de pomme de terre, et baptisée EUROPP – Efficiency in the Use of Resources: Optimization in Potato Production.

Ce manuel explique dans le chapitre 1 les rôles de l'eau et de l'azote dans la croissance et le développement de la pomme de terre et les chapitres suivants, présentent une évaluation des procédures disponibles pour arriver à un équilibre correct entre intrants et production. Le chapitre 2 décrit les tendances du statut en azote de la culture (azote total et nitrate), comment ce statut peut être déterminé par des méthodes destructives et non-destructives, comment échantillonner la culture de manière optimale, et quelle est l'utilité de telles mesures en matière d'aide à la décision pour l'application d'azote. De manière similaire, le chapitre 3 s'attache aux questions conçernant le statut en azote du sol, en considérant les sources en azote organique et minéral, et montre l'importance d'un échantillonnage adéquat. Il explique l'importance de la nitrification et de la dénitrification ainsi que les mécanismes des pertes en azote. Le chapitre 4 explique le rôle de l'eau dans la plante et dans le sol, comment le statut en eau peut être mesuré dans chacun des cas, et comment ces mesures peuvent être utilisées en pratique. Le Chapitre 5 passe en revue les systèmes d'aide à la décision existants pour la fourniture en azote et en eau, quelles sources d'azote peuvent être utilisées (amendements organiques ou engrais manufacturés), quelles techniques d'irrigation sont adéquates, et enfin quels systèmes d'aide à la décision sont disponibles pour l'eau et l'azote. Les différences importantes entre une production de pomme de terre en conditions irriguées et une production tributaire des pluies sont également considérées. Le chapitre 6 explore les développements possibles pour le futur: Quel est le rôle de la modélisation de la croissance de la culture et de la simulation sur ordinateur? Quels sont les risques et les leçons à tirer des tendances dans les pratiques actuelles? Que savons-nous sur les conséquences de l'agriculture biologique? Quelles sont les opportunités de l'agriculture de précision? Et que devons nous encore apprendre? Le livre conclut avec les perspectives des développements récents, en ce compris ceux issus de l'action concertée.

Le manuel actualise les connaissances d'une manière simple et facilement compréhensible. Il s'adresse aux agriculteurs, aux services de consultance, aux chercheurs et aux décideurs dans les secteurs de l'agriculture, de la recherche et de la politique.

Stickstoff- und Wassereinsatz im Kartoffelbau

Zusammenfassung in Deutsch übersetzt von Dr. N.U. Haase

Stickstoff (N) und Wasser zählen zu den wichtigsten Produktionsfaktoren bei der Kartoffelerzeugung. Das Wechselspiel zwischen Zufuhr und Freisetzung steuert in hohem Maße die Effizienz der Produktion, da sowohl Ertrag als auch Qualität darauf empfindlich reagieren können. Sowohl Überschuß als auch Mangel führen zu Probleme. Hinzu kommt, daß ein Überschuß auch negative Auswirkungen auf die Umwelt hat. Im Falle der künstlichen Bewässerung bedeutet dieses, daß die (oft) endlichen Resourcen nicht effizient genutzt werden. Hinzu kommt, daß Auswaschungsverluste bei den Nährstoffen auftreten können. Ein Zuviel an Stickstoff führt ebenfalls zu einer Auswaschung in das Grundwasser (i.d.R. in Form von Nitrat). Nicht zuletzt deshalb können beide Produktionsfaktoren einer staatlichen Reglementierung unterzogen werden.

Dieses Buch entstand im Rahmen einer konzertierten Aktion der Europäischen Kommission mit der Maßgabe, die effizienten Nutzung von Stickstoff und Wasser im Kartoffelbau zu thematisieren (EUROPP- Efficiency in the Use of Resources: Optimisation in Potato Production; EUROPP- Effiziente Resourcennutzung: Optimierung der Kartoffelproduktion).

In Kapitel 1 des Handbuches wird zunächst auf die Bedeutung von Wasser und Stickstoff beim Wachstum und bei der Entwicklung der Kartoffelpflanze eingegangen. Die weiteren Kapitel diskutieren dann verschiedene Verfahren, um eine ausgeglichene Versorgung der Pflanzen zu erreichen. So vertieft Kapitel 2 das Thema des Stickstoffhaushaltes (Gesamt-N und Nitrat), inwieweit der Versorgungsgrad durch invasive und nicht-invasive Methoden erfasst werden kann, wie eine korrekte Probenahme zu erfolgen hat, und wie derartige Ergebnisse im Rahmen einer Entscheidungsfindung für die Stickstoffausbringung zu nutzen sind. Der Stickstoffhaushalt des Bodens wird in Kapitel 3 behandelt. Dabei werden sowohl die anorganischen als auch die organischen N-Quellen angesprochen. Aber auch die Bedeutung einer korrekten Probenahme kommt nicht zu kurz. Des weiteren werden die Vorgänge der Nitrifizierung und der Denitrifizierung behandelt und mögliche Verluste aufgezeigt. Kapitel 4 beschreibt die Rolle des Wassers in Pflanze und Boden. Angesprochen wird, wie der jeweilige Wasserstatus bestimmt und wie diese Messungen in der Praxis genutzt werden können. Kapitel 5 erörtert, welche Stickstoffquellen genutzt werden können (hofeigene und zugekaufte

Dünger), und welche Bewässerungssysteme geeignet sind. Des weiteren wird ein Überblick über verfügbare Entscheidungshilfen zur Stickstoff- und Wasserzufuhr gegeben. Auch werden die Unterschiede zwischen künstlicher Bewässerung und natürlicher Wasserversorgung angesprochen. Kapitel 6 widmet sich den zukünftigen Entwicklungen: Welche Rolle werden Wachstumsmodelle und die Computer-Simulation spielen? Worin liegen die Risiken bei Fortführung der gegenwärtigen Praxis? Was wissen wir über die Auswirkungen der organischen Bewirtschaftungsweise? Welche Vorteile hat die integrierte Wirtschaftsweise, und in welchen Bereichen gibt es noch Wissenslücken? Das Buch endet mit Perspektiven jüngster Entwicklungen einschließlich einesFazits aus dieser konzertierten Aktion.

Insgesamt soll das Buch dazu dienen, vorhandenes Wissen in einfacher und leicht verständlicher Form auf den neuesten Stand zu bringen. Zielgruppen sind Landwirte sowie Berater, Forscher und Entscheidungsträger aus den Bereichen Landwirtschaft, Forschung und Politik.

Χορήγηση αζώτου και νερού στην καλλιέργεια της πατάτας

Greek summary translated by Prof. Panagiotopoulos

Ο εφοδιασμός των φυτών με νερό και άζωτο αποτελεί βασικό πρόβλημα στην καλλιέργεια της πατάτας. Υπάρχει ισορροπία μεταξύ των ποσοτήτων που εφαρμόζονται και εκείνων που συγκομίζονται που μπορούν έντονα να επηρεάσουν το οικονομικό όφελος της καλλιέργειας, επειδή τόσο η παραγωγή, όσο και η ποιότητα των κονδύλων εξαρτάται σε μεγάλο βαθμό από το σωστό εφοδιασμό με νερό και άζωτο. Υπερβολικές ποσότητες και τροφοπενίες των παραγόντων αυτών δημιουργούν προβλήματα στην καλλιέργεια. Επιπλέον, υπερβολικές ποσότητες έχουν σαν αποτέλεσμα την δημιουργία περιβαλλοντολογικών προβλημάτων. Συγκεκριμένα, στην περίπτωση του νερού, που είναι ένας φυσικός πόρος , υπο διαρκή έλλειψη χρησιμοποιείται αναποτελεσματικά και μπορεί να συμβάλλει στην έκπλυση. Στις περιπτώσεις που εφαρμόζονται μεγάλες ποσότητες αζώτου, η περίσσεια είναι βέβαιο ότι θα εκπλυθεί στα υπόγεια νερά. Εξαιτίας αυτών των πιθανών αρνητικών επιπτώσεων, η εφαρμογή αζώτου και η άντληση νερού αποτελούν αντικείμενα κανονισμών των Κρατικών Υπηρεσιών. Αυτό το βιβλίο εκδόθηκε στα πλαίσια μιας Συντονισμένης Δράσης που χρηματοδοτείται από την Ευρωπαική Ενωση με σκοπό την αποτελεσματική χρήση αζώτου και νερού στην καλλιέργεια πατάτας και ονομάζεται EUROPP - Efficiency in the Use of Resources: Optimisation in Potato Production (Αποτελεσματικότητα στη χρησιμοποίηση των φυσικών πόρων: Βελτιστοποίηση στην καλλιέργεια πατάτας).

Αυτός ο οδηγός, στο κεφάλαιο 1 εξηγεί τους ρόλους του νερού και του αζώτου στην αύξηση και εξέλιξη των πατατοφύτων, ενώ στα υπόλοιπα κεφάλαια παρουσιάζεται μια εκτίμηση των διαθέσιμων τεχνικών για την επίτευξη του σωστού ισοσκελισμού μεταξύ εισροών και παραγωγής. Το κεφάλαιο 2 πραγματεύεται την περιεκτικότητα των φυτών σε άζωτο (ολικό και νιτρικό) και συγκεκριμένα τις μεθόδους προσδιορισμού σε νωπούς ιστούς ή μετά από καύση, τον τρόπο δειγματοληψίας, και τη σπουδαιότητα των μετρήσεων αυτών για τον υπολογισμό της αζωτούχου λιπάνσης. Αντίστοιχα στο κεφάλαιο 3, τίθεται το ερώτημα για την περιεκτικότητα του εδάφους σε άζωτο, λαμβάνοντας υπόψιν και τις δυό πηγές του αζώτου, ανόργανο και οργανικό, και παράλληλα, αναλύεται η σπουδαιότητα της επαρκούς δειγματοληψίας. Επιπλέον, εξηγείται η σπουδαιότητα της νιτροποίησης και της απονιτροποίησης και υπό ποιές συνθήκες γίνονται απώλειες αζώτου. Στο κεφάλαιο 4 αναλύεται ο ρόλος του νερού στα φυτά και το έδαφος, αναφέρονται οι μέθοδοι προσδιορισμού του νερού σε αυτά και, συγχρόνως αποσαφηνίζεται πώς οι μετρήσεις αυτές μπορούν να αξιοποιηθούν στην πράξη. Στο κεφάλαιο 5 γίνεται ανακεφαλαίωση των υπαρχόντων συστημάτων για την υποβοήθηση στην λήψη αποφάσεων για τον εφοδιασμό με άζωτο και νερό, καθώς και των πηγών αζώτου, που μπορούν να χρησιμοποιηθούν (οργανικά λιπάσματα και λιπάσματα σε σάκους). Επίσης, εξηγείται ποιές αρδευτικές πρακτικές είναι κατάλληλες, και ποιά συστήματα για την υποβοήθηση στην λήψη αποφάσεων σχετικά με το άζωτο και το νερό είναι διαθέσιμα. Ακολούθως, εξετάζονται οι σημαντικές διαφορές ανάμεσα στις αρδευόμενες καλλιέργειες και σ' εκείνες που δέχονται μόνο βρόχινο νερό. Στο κεφάλαιο 6 διερευνώνται πιθανές μελλοντικές εξελίξεις. Ποιός είναι ο ρόλος της δημιουργίας μοντέλων για την ανάπτυξη των καλλιεργειών και προσομοιωτών μέσω ηλεκτρονικού υπολογιστή. Ποιοί είναι οι κίνδυνοι και ποιά τα συμπεράσματα από τις επικρατούσες καλλιεργητικές τεχνικές. Τί γνωρίζουμε για τις συνέπειες της "οργανικής" γεωργίας. Ποιές είναι οι δυνατότητες της χρησιμοποίησης γεωργικών μηχανημάτων ακριβείας και τί επιπλέον πρέπει να μάθουμε. Το βιβλίο τελειώνει με την παράθεση των τελευταίων δεδομένων συμπεριλανβανομένων εκείνων που προήλθαν από τη συγκεκριμένη Συντονισμένη Δράση.

Το βιβλίο ενημερώνει με έναν απλό, εύκολο και κατανοητό τρόπο και απευθύνεται σε αγρότες, σύμβουλους των αγροτών, ερευνητές και κέντρα λήψεως αποφάσεων για τους αγρότες, τους ερευνητές και τους πολιτικούς.

Gestione dell'Acqua e dell'Azoto nella produzione della patata

Riassunto in Italiano tradotto da Luisa Dalla Costa

Le disponibilità di acqua e di azoto hanno un ruolo chiave per la produzione nella coltura di patata. Esiste un equilibrio tra la loro disponibilità e la conseguente produzione che determina la redditività della coltura, in quanto condiziona sia l'aspetto quantitativo della produzione, che quello qualitativo, entrambi legati ad una corretta gestione di acqua e azoto. Il loro eccesso come la loro carenza causano problemi alla coltura. Gli eccessi, inoltre, possono provocare danni ambientali: nel caso dell'acqua ciò implica che un risorsa limitata viene usata in modo inefficiente e verrà lisciviata. Nel caso di troppo azoto, l'eccesso verrà dilavato e perduto nella falda. Per questi motivi la dose di azoto distribuita e l'estrazione dell'acqua dalla falda freatica possono essere regolamentate dalle autorità governative. Questo libro è stato realizzato nel corso di un'Azione Concertata finanziata dalla Comunità Europea sull'uso efficiente di azoto e acqua nella coltura di patata, chiamato EUROPP - Efficienza dell'Uso delle Risorse: Ottimizzazione nella Produzione della Patata.

Il manuale spiega il ruolo dell'acqua e dell'azoto nella crescita e nello sviluppo della coltura di patata nel capitolo 1, e, nei capitoli successivi, presenta le procedure disponibili per stilare ad un bilancio corretto tra gli input e la produzione. Il capitolo 2 illustra gli andamenti del contenuto di azoto nella pianta (N totale e nitrati), come esso possa essere determinato con metodi invasivi e non, come prelevare i campioni, qual è il ruolo di queste misure nel guidare decisioni sulle applicazioni di azoto. Analogamente, il cap. 3 si sofferma sulla disponibilità di N nel suolo, considerando la provenienza organica e inorganica, e illustra l'importanza di un campionamento adeguato del terreno. Spiega inoltre l'importanza del processo di nitrificazione e denitrificazione, e come avvengono le perdite. Il cap. 4 spiega il ruolo dell'acqua nella pianta e nel terreno, come si possa misurare lo stato dell'acqua in questi due comparti, e come le misure possono essere utilizzate nella pratica. Il cap. 5 analizza i sistemi di supporto alle decisioni (DSS) esistenti per l'azoto e l'acqua, quali fonti di azoto possano essere impiegate (concimi organici e fertilizzanti chimici), quali pratiche irrigue sono adatte, e quali sistemi di supporto sono disponibili per acqua e azoto. Analizza inoltre l'importante differenza che esiste tra coltura irrigua e coltura in asciutta. Il capitolo 6 esplora i possibili sviluppi futuri: qual è il ruolo della modellizzazione e della simulazione tramite computer? Quali sono i rischi e gli errori delle pratiche attuali? Cosa sappiamo delle conseguenze dell'agricoltura biologica? Che possibilità offre l'agricoltura di precisione? Quali lati sono ancora oscuri e hanno bisogno di ulteriori ricerche? Il libro chiude la panoramica con le prospettive dei recenti sviluppi, incluse le conclusioni cui è giunta l'azione concertata.

Il libro aggiorna le conoscenze sull'argomento in modo semplice e facilmente comprensibile, ed è destinato ad agricoltori, assistenti tecnici, ricercatori e personale amministrativo che devono operare scelte a livello aziendale, di ricerca, e a livello politico.

Gospodarowanie azotem i wodą w produkcji ziemniaka

Tłumaczenie streszczenia na jezyk Polski: Alicja Pecio

Kluczowym zagadnieniem w produkcji ziemniaka jest odpowiednie zaopatrzenie roślin w azot i wodę . Obydwa te czynniki wpływają istotnie na wielkość i jakość plonu, a zachowanie właściwych proporcji pomiędzy wielkością nakładów i przyrostem plonu decyduje o opłacalności produkcji. Niekorzystny wpływ na plonowanie ziemniaka wywiera zarówno nadmiar jak i niedobór azotu oraz wody. Stosowanie ich w nadmiarze jest szkodliwe również ze względów środowiskowych. Nawadnianie zbyt wysokimi dawkami oznacza nieefektywne wykorzystanie ograniczonych zasobów wodnych, a ponadto powoduje wymywanie azotu z gleby do wód gruntowych. Ze względu na możliwość występowania tych niekorzystnych zjawisk, nawożenie azotem i nawadnianie powinny być przedmiotem regulacji prawnych. Książka powstała w ramach finansowanego przez Unię Europejską programu *EUROPP*, poświęconego zagadnieniom efektywnego wykorzystania azotu i wody w produkcji ziemniaka.

W 1 rozdziale podręcznika omówiono wpływ wody i azotu na wzrost i rozwój roślin. W dalszych rozdziałach zaprezentowano sposoby zwiększania efektywności nawożenia i nawadniania. W rozdziale 2 przedstawiono destrukcyjne i niedestrukcyjne metody oceny stanu odżywienia roślin azotem (N ogólny i azotanowy), omówiono metodykę pobierania prób oraz możliwości optymalizacji nawożenia azotem na podstawie analizy roślin. Rozdział 3 poświęcono zagadnieniom azotu glebowego: źródła azotu organicznego i mineralnego, przestrzenna zmienność zawartości N w glebie i metodyka pobierania prób. Omówiono procesy nifryfikacji i denitryfikacji oraz zagadnienie strat azotu glebowego. Rozdział 4 traktuje o roli wody w roślinie i glebie, metodach jej pomiaru i o możliwościach praktycznego wykorzystania tego rodzaju pomiarów. W rozdziale 5 zaprezentowano dostępne systemy doradztwa w zakresie nawożenia azotem i nawadniania. Omówiono także istotne różnice pomiędzy nawożeniem ziemniaka uprawianego w warunkach nawadniania i bez nawadniania. W rozdziale 6 rozważono perspektywiczne kierunki rozwoju nauki i doradztwa w zakresie uprawy ziemniaka: modelowanie i komputerowa symulacja wzrostu i rozwoju roślin, skutki gospodarki „organicznej", możliwości jakie stwarza rolnictwo precyzyjne. Podsumowano osiągnięcia programu EUROPP i wskazano luki w wiedzy dotyczącej uprawy ziemniaka.

Książka prezentuje wiedzę w sposób prosty i dostępny. Adresowana jest do rolników, doradców, pracowników naukowych i kół decydenckich, kształtujących politykę rolną.

Gestão do azoto e da água na produção de batata

Resumo em Português traduzido por F. Martins

A rega e a adubação azotada são factores determinantes na produção de batata, dado que a produtividade e a qualidade dos tubérculos estão muito dependentes de um correcto fornecimento de água e azoto. Se fornecidos por defeito ou por excesso ocasionam problemas na cultura e, no caso particular de excesso, podem conduzir a problemas ambientais. Na rega, quando é usada muita água, podem ocorrer perdas por infiltração profunda, reduzindo a eficiência de aplicação de um bem cada vez mais limitado. De igual forma, na adubação azotada em excesso, a parte do azoto não utilizada pela cultura é lixiviada, contaminando as águas subterrâneas. É devido a estes efeitos adversos potenciais que a utilização de azoto e da água podem ser sujeitas a regulamentação especial por parte das autoridades competentes.

Tendo em conta o anteriormente referido, o presente livro foi produzido durante a Acção Concertada sobre o uso eficiente do azoto e da água na produção de batata, designada por *EUROPP – Efficiency in the Use of Resources: Optimisation in Potato Production* (Eficiência no Uso de Recursos: Optimização na Produção de Batata), e financiada pela União Europeia.

O Manual aborda, no Capítulo 1, as funções da água e do azoto no crescimento e desenvolvimento da cultura e apresenta uma avaliação dos métodos disponíveis para alcançar um balanço correcto entre a produção de batata e o fornecimento de água e azoto. No Capítulo 2 são abordadas as questões do estado nutricional da planta em azoto (total e sob a forma de nitrato), as técnicas como este estado nutricional pode ser avaliado (usando métodos invasivos ou não invasivos), a amostragem mais correcta para o determinar e a forma em que aquelas técnicas podem servir de base à decisão de aplicar azoto. De um modo similar, o Capítulo 3 aborda as questões do azoto no solo, considerando as formas orgânicas e inorgânicas, explica a importância dos processos de nitrificação, desnitrificação, como ocorrem as perdas deste nutrientes e realça a importância de uma amostragem adequada para avaliação da disponibilidade em azoto. O Capítulo 4 explica as funções da água nas plantas e no solo, a forma de medir a quantidade existente em cada um deles, e como é que estas medidas podem ser usadas na prática. No Capítulo 5 são revistos os sistemas de tomada de decisão existentes para a adubação azotada e para a rega, quais as fontes de azoto que podem ser usadas (fertilizantes orgânicos e/ou minerais), quais os métodos de rega aconselháveis e quais os sistemas de tomada de decisão utilizáveis para a rega e para a adubação azotada. São também abordadas as diferenças entre os sistemas de produção de batata com e sem rega. O Capítulo 6 aborda potenciais linhas de desenvolvimento e interesse futuro: Qual a função da modelação das culturas e dos modelos de simulação? Quais os riscos e os ensinamentos das práticas correntes? Que consequências da designada 'agricultura biológica'? Quais as oportunidades para a agricultura de precisão? Quais as necessidades de conhecimento para o futuro? O livro conclui com as perspectivas mais recentes em diferentes domínios, incluindo os provenientes da Acção Concertada.

Em resumo, presente livro actualiza, de uma forma simples e facilmente compreensível, o conhecimento nos domínios referidos e destina-se a ser utilizado por agricultores, extensionistas, investigadores e decisores ao nível da produção, da investigação e da política sectorial.

Manejo del nitrógeno y del agua en la producción de patata

Resumen al Espagnol traducido por A. Borruey

El riego y la fertilización nitrogenada son factores determinantes para la producción de patata, dado que la productividad y la calidad de los tubérculos dependen en gran medida de un aporte correcto de agua y nitrógeno. Tanto el aporte deficitario como el excesivo ocasionan problemas en el cultivo y, en el caso concreto del exceso, se pueden originar problemas medioambientales. En el riego, cuando se aporta mucha agua, llegan a producirse pérdidas por infiltración profunda, lo que reduce la eficiencia en el uso de un bien cada vez mas limitado. De igual modo la fertilización nitrogenada excesiva hace que el nitrógeno se lixivie, contaminando las aguas subterráneas. A causa de estos potenciales efectos adversos, la utilización del nitrógeno y del agua pueden ser objeto de una regulación especial por parte de las autoridades competentes.

Este libro, como respuesta a lo anteriormente expuesto, se elaboró en el curso de una Acción Concertada financiada por la Unión Europea sobre el uso eficiente del nitrógeno y del agua en la producción de patata, denominada EUROPP - Efficiency in the use of Resources; Optimisation in Potato Production (Eficiencia en el uso de Recursos: Optimización en la Producción de Patatas).

Este manual aborda en el Capítulo 1 las funciones del agua y del nitrógeno en el crecimiento y desarrollo del cultivo y, en los capítulos siguientes presenta una evaluación de los procedimientos disponibles para conseguir un correcto equilibrio entre la producción de patata y el aporte de agua y nitrógeno. El Capítulo 2 estudia cuestiones sobre el estado nutricional de la planta en nitrógeno (total y bajo la forma de nitrato), las técnicas para evaluar este estado nutricional (usando métodos invasivos y no invasivos), la técnica de muestreo mas correcta y la forma en que estas técnicas pueden servir de base para la aplicación del nitrógeno. De un modo similar, el Capítulo 3 plantea cuestiones sobre la situación del nitrógeno en el suelo, considerando las formas orgánicas e inorgánicas, explica la importancia de los procesos de nitrificación, desnitrificación, cómo se producen las pérdidas de este nutriente y resalta la importancia de una correcta toma de muestras para valorar la disponibilidad de nitrógeno. El Capítulo 4 expone las funciones del agua en las plantas y en el suelo, la forma de medir la cantidad existente en cada medio, y cómo pueden usarse estas mediciones en la práctica. En el Capítulo 5 se revisan los sistemas de toma de decisión existentes para el abonado nitrogenado y para el riego, qué fuentes de nitrógeno pueden usarse (fertilizantes orgánicos y/o minerales), qué métodos de riego son aconsejables y qué sistemas de toma de decisión son utilizables para el riego y para la fertilización nitrogenada. Se abordan también las diferencias entre los sistemas de producción de patata con y sin riego. El Capítulo 6 indaga sobre potenciales líneas de progreso futuro: ¿cuál es el papel de la planificación del cultivo y de la simulación por ordenador? ¿cuales son los riesgos y enseñanzas de las prácticas de cultivo corrientes? ¿ qué sabemos sobre las consecuencias de la "agricultura biológica"? ¿Cuales son las oportunidades para la agricultura altamente tecnificada? y ¿ qué precisamos aprender todavía?. El libro concluye con las perspectivas de los recientes progresos, incluidos los procedentes de la Acción Concertada.

En resumen, el presente libro actualiza información sobre los temas referidos de un modo simple y fácilmente comprensible, estando dirigido a agricultores, técnicos agrícolas, investigadores y responsables a nivel de producción, investigación y política sectorial agraria.

Subject index

Printed in the United States
by Baker & Taylor Publisher Services